SpringerBriefs in Research and Innovation Governance

More information about this series at http://www.springer.com/series/13811

Konstantinos Iatridis · Doris Schroeder

Responsible Research and Innovation in Industry

The Case for Corporate Responsibility Tools

 Springer

Konstantinos Iatridis
School of Management
University of Bath
Bath
UK

Doris Schroeder
College of Health
University of Central Lancashire
Preston
UK

ISSN 2452-0519   ISSN 2452-0527   (electronic)
SpringerBriefs in Research and Innovation Governance
ISBN 978-3-319-21692-8   ISBN 978-3-319-21693-5   (eBook)
DOI 10.1007/978-3-319-21693-5

Library of Congress Control Number: 2015946103

Springer Cham Heidelberg New York Dordrecht London

Printed on acid-free paper

Springer International Publishing AG Switzerland is part of Springer Science+Business Media
(www.springer.com)

To Eleni and Myrto
To Armin

Foreword

Businesses are responsible. They are responsible in different ways and for different things. They have legal responsibilities, contractual responsibilities and responsibilities towards their shareholders, customers and employees. All of these responsibilities have a component of moral responsibility, of doing the right thing for the right reasons. In addition to these responsibilities of individual companies, the overall economic system is responsible for providing the goods and services that a society requires.

In market economies, the foundational idea is that the moral responsibility for the overall outcome of economic activity is not attributed to the individual company or human being, but that the system allows all participants to maximize their own interests, which leads to the optimal distribution of goods and use of resources. This is what Adam Smith (2008, p. 119) meant when he said: 'It is not from the benevolence of the butcher, the brewer, or the baker that we expect our dinner, but from their regard to their own interest. We address ourselves, not to their humanity but to their self-love, and never talk to them of our own necessities but of their advantages'.

In practice, this idea often fails. Companies, and the individuals within them, can be irresponsible and deliberately disregard their responsibilities. Newspapers are full of reports about irresponsible behaviour by and within companies. More importantly, responsibilities may be unclear and contradictory. A company's responsibility to maximize shareholder wealth does not always sit easily with its responsibility for the natural environment or its responsibilities as a corporate citizen existing within a society.

One response to this complex of interlinking responsibilities has been the development of the idea and practice of corporate responsibilities. Positing such responsibilities is an important first step, but it has long been recognized that they need to be rendered practically relevant and implemented. The discourse around corporate responsibility has taken up this challenge and developed a number of tools and methods to put these responsibilities in practice.

One novel type of responsibility that is gathering increasing interest and growing in relevance is that of responsible research and innovation (RRI). The idea behind this concept is that both the processes and the outcomes of research and innovation activities should be acceptable and socially desirable (van den Hoven 2013; Owen et al. 2013; von Schomberg 2013). The concept itself is novel, but it builds on a significant number of recognized activities, including technology ethics, technology assessment, science and technology studies, and research policy.

RRI raises two challenges of particular importance for this book: the question of practical relevance and the role of industry in RRI. Much of the work in RRI focuses on research and innovation activities that are publicly funded and typically undertaken in universities or public research organizations. This is understandable because the discourse has been promoted by public research funders such as the European Commission under its research framework Horizon 2020, the Netherlands Organisation for Scientific Research, the Research Council of Norway and the UK Engineering and Physical Sciences Research Council, to name but a few. However, publicly funded research tends to take place at an early technology readiness level, which can make it very difficult to predict eventual outcomes and social consequences. Most research and development activities that are closer to market and therefore likely to have an immediate impact on end users are undertaken by private companies. Recommendations and guidelines aimed at publicly funded research and innovation are unlikely to resonate with such companies. For RRI to be practically relevant to private companies, their interests and incentives need to be taken into account.

If RRI is to contribute to the acceptability and desirability of research processes and products, one first has to ask how this can be achieved. Or, put differently, how does RRI go beyond its predecessor or component activities like technology assessment in achieving a desired outcome? One promising response to this question is to position RRI at a different level from that of the multiplicity of existing responsibilities. RRI does not simply add more responsibilities to the crowded field, but should be seen as a higher level of responsibility, a meta-responsibility (Stahl 2013). This means that the novelty of RRI is that it aims to shape, maintain, develop, coordinate and align existing responsibilities. We do not need to invent new responsibilities, but should rather make sure that the existing ones work in a way that leads to an overall desirable outcome.

Doris Schroeder and Konstantinos Iatridis make a contribution to addressing the two challenges of RRI. They provide an overview of established tools of corporate responsibility, thus clarifying which responsibilities already exist and how they can be implemented. Moreover, they show how these tools can be used for the purposes of RRI. This allows decision-makers to adopt RRI and accept the higher-level responsibility of ensuring that their research and innovation activities are consistent with the overall outlook and responsibilities of the organization.

Iatridis and Schroeder understand that the value of RRI is in its practical relevance. They explain the relationship between the tools of corporate responsibility and RRI by providing a set of case studies. They furthermore offer the reader a dialogical tool to reflect on their current levels of responsibilities which will allow them to identify current strengths and weaknesses.

In the twenty-first century, reinventing the wheel in corporate governance is unattractive, as most businesses are already overwhelmed with the choices presented by business analysts, from innovation strategy to ethical conduct. A book that applies existing tools (corporate responsibility) to a new field (RRI) is therefore highly welcome. It may eventually allow businesses to achieve their own goals *and* contribute to the greater good of society.

Bernd Stahl

References

Owen R, Bessant J, Heintz M (eds) (2013) Responsible innovation: managing the responsible emergence of science and innovation in society. Wiley, London

Smith A (2008) An inquiry into the nature and causes of the wealth of nations [Bullock edition]. Wildside Press LLC, Rockville

Stahl BC (2013). Responsible research and innovation: the role of privacy in an emerging framework. Sci Public Policy 40(6):708–716. http://doi.org/10.1093/scipol/sct067

van den Hoven J, Jacob K, Nielsen L, Roure F, Rudze L, Stilgoe J (2013) Options for strengthening responsible research and innovation. Directorate-General for Research and Innovation, European Commission, Brussels

von Schomberg R (2013) A vision of responsible research and innovation. In: Owen R, Bessant J, Heintz M (eds) Responsible innovation: managing the responsible emergence of science and innovation in society. Wiley, London, pp 51–74

Pre-publication Endorsements

Acknowledgments

As academics who previously worked in industry, we enjoyed writing this book. Our sincere hope is that it will be useful to industry *and* academia.

Funding for the book was obtained from the European Commission through the 'Responsible Industry' project.[1]

In addition to expressing our appreciation for the funding, we would also like to acknowledge the spirited and knowledgeable contribution that Dr. Karen Fabbri (SwafS) made to the whole community of researchers working on responsible research and innovation (RRI) with EU funding. As the policy officer of five RRI projects, including Responsible Industry, she managed to infuse our interactions with energy, creativity and sparkle.

For one of us, Doris Schroeder, this is the third collaboration with Fritz Schmuhl, the Senior Editor at Springer, whom we very warmly commend on his engagement, his responsiveness and his interest in books. Likewise, it is the third collaboration with a fantastic copy editor, Paul Wise, from Cape Town. Thanks to both!

Thanks also to Julie Cook Lucas for insightful comments on an early draft; to Kate Chatfield and Dr. Nearchos Paspallis, our partners in the UCLan Cyprus dream team; to Dr. Lee Chatfield for inviting Doris to UCLan Cyprus and to Dr. Nigel Harrison for funding the initial secondment; to the Responsible Industry team, and especially Prof. Bernd Stahl, for input into our work; to Prof. Richard Owen for knowledgeable discussions on social innovations in Brussels, which we will have to leave for another occasion; to Drs Nigel Gericke in South Africa and Thomas Frenken in Germany for sharing information about their innovative products, which are featured in Chap. 2; and to Dr. Lino Paula, who set the Responsible Industry project off on a sound course.

[1]The research leading to this book has received funding from the European Community's Seventh Framework Programme (FP7/2007–2013) under Grant Agreement No. 609817 (http://www.responsible-industry.eu).

Contents

Abbreviations

BoP	Bottom of pyramid
BSCI	Business social compliance initiative
BSI	British Standards Institution
CERES	Coalition for environmentally responsible economies
CR	Corporate responsibility
CRT	Caux round table principles for business
DSM	Diagnostic and statistical manual of mental disorders
EMAS	Eco-management and audit scheme
ETI	Ethical trading initiative
FTA	Foreign Trade Association
GDP	Gross domestic product
GMOs	Genetically modified organisms
GRI	Global reporting initiative
HLL	Hindustan Lever limited
IEC	International Electrotechnical Commission
ILO	International Labour Organization
ISEA	Institute of Social and Ethical Accountability (also known as AccountAbility)
ISEAL Alliance	International Social and Environmental Accreditation and Labelling Alliance
ISO	International Organization for Standardization
MNE	Multinational enterprises
NGOs	Nongovernmental organizations
OECD	Organisation for Economic Co-operation and Development
RRI	Responsible research and innovation
RUCs	Random unannounced checks
SAAS	Social accountability accreditation services
SAI	Social Accountability International
SASC	South African San Council
SwafS	Science with and for society

ToP	Top of pyramid
UNEP	United Nations Environment Programme
UNGC	United Nations Global Compact

Chapter 1
Introduction: More Responsible Researchers and Innovators?

Abstract This book argues that corporate responsibility (CR) tools can help realize the goals of the responsible research and innovation (RRI) framework. RRI is a newly emerging governance framework, promoted by public funders of research such as the European Commission. When RRI is applied in industry, funder requirements are not enough to implement it, given that private research and innovation funds are involved. Instead, we argue, CR tools are well suited to bridge this gap as they are self-regulatory and have been established to promote a common understanding as well as a common means of performance evaluation globally.

Keywords Responsible research and innovation · Corporate responsibility

Innovators and researchers want to achieve progress, profits or both. Progress comes through new knowledge, and profits through new products and services. The two richest billionaires in the world, Bill Gates and Carlos Slim Helu (Forbes n.d.), made their fortunes with innovations that were unthinkable 150 years ago: computers and telecommunications.

The accelerated speed of research and innovation (RRI) has brought many benefits to humankind. To give just one example, today childhood leukaemia has a survival rate of over 90 % (Simon 2012) for those with access to health care. At the same time, the speed of economic growth facilitated through innovation has brought many challenges. For instance, climate change 'is the major, overriding environmental issue of our time', according to the United Nations Environment Programme (UNEP n.d.); a challenge caused by the burning of fossil fuels which is necessary for industrial production, according to NASA (NASA n.d.).

Ideally, *responsible* researchers and innovators achieve progress (and profits) without damaging the prospects of current and future populations. Currently, however, this is not happening. For instance, it is estimated that due to climate change alone, 25 % of all species living on land will be lost by 2050 (CHGE n.d.).

There have been countless suggestions as to how innovators, researchers and businesses could both respond to humanity's challenges and be more responsible

K. Iatridis and D. Schroeder, *Responsible Research and Innovation in Industry*,
SpringerBriefs in Research and Innovation Governance,
DOI 10.1007/978-3-319-21693-5_1

in their work. These range from asking innovators to 'put the genie back into the bottle', for instance in the context of genetic modification (Ishii-Eiteman 2013), to suggestions that science will save the world, especially through genetic modification (Naam 2013). In this book, we are not going to suggest anything as dramatic at either end of the spectrum. Instead, we will:

1. familiarize the reader with the relatively new concept of responsible research and innovation (RRI) (see Chap. 2);
2. remind the reader of the essentials of the concept of corporate responsibility (CR) (see Chap. 3); and
3. analyse how tools developed in the context of CR can be used to monitor and guide RRI performance (see Chaps. 4 and 5).

1.1 Why Corporate Responsibility Tools in Responsible Research and Innovation?

Why do we think that CR tools can be used to monitor and guide RRI performance? RRI is a relatively new concept in innovation governance. While it includes elements of earlier governance frameworks (e.g. technology assessment to avoid significant risks to human health and the environment), it has taken a significant integrative step. Armin Grunwald comments:

> Researchers and practitioners from the fields of technology assessment, engineering ethics, sustainability research, science and technology studies, risk assessment, technology foresight and strategic management are becoming part of a new and emerging research community: a community, which discusses issues of Responsible Research and Innovation (RRI) in an integrative manner (Grunwald 2014, p. 1).

The development of tools and metrics usually *follows* the development of a concept. For instance, the European Commission has issued a tender specifically designed to develop metrics for RRI, and the successful group will only report its major findings around 2017 (European Commission 2013). This is an indication that tools and metrics for RRI are currently rare or still under development. However, it is also important to note that effective governance frameworks have to fulfil certain criteria. As we will explain in more detail in Chap. 4, one of these criteria is that governance tools should not significantly overlap with existing tools but should complement them. It is therefore highly important to see which current innovation governance frameworks can be used effectively in moving RRI forward.

This book argues that existing CR tools can assist in realizing the goals of the emerging RRI framework.

In a globalized context, it is more and more difficult for national governments to monitor and guide corporate conduct. As a result, responsibilities are moved increasingly to the private sector or into joint ventures. Such responsibilities can align closely with the imperative to 'do no harm', for instance in terms of avoiding bribery or environmental damage. Or they can—which aligns well with new elements in RRI—give corporations a co-responsibility to 'do good' beyond corporate philanthropy.

For instance, access to health care is a fundamental human right which has been part of international law since 1948 (United Nations 1948, Art. 25). National governments are usually regarded as the primary duty-bearers tasked with ensuring that human rights are respected, protected and fulfilled. However, more recently it has been argued that pharmaceutical companies have a co-responsibility to fulfil the human right to health (Schroeder 2011). This is most obvious, at the global scale, in United Nations Millennium Goal 8 Target E: 'In cooperation with pharmaceutical companies, provide access to affordable essential drugs in developing countries' (United Nations n.d.).

CR tools are well suited to a globalized world, where corporations are given more and more responsibilities. They are self-regulatory tools that have been established to promote a common understanding as well as a common means of performance evaluation globally. Today, companies with high public acceptance levels and a good reputation as corporate citizens face challenges far beyond increasing and sustaining their profits through new innovations: they are required by state and societal actors to play a more socially active role (Porter and Kramer 2011). Self-regulation plays a crucial role in helping companies do so.

CR tools are formal technical documents, issued under the auspices of international organizations such as the International Organization for Standardization (ISO), that establish criteria, methods, processes and practices intended to help companies improve their internal organization and operational efficiency. Importantly, in the context of RRI, which has a strong engagement element (see Chap. 2), CR tools are often the product of a wide multistage consultation process in which various groups of stakeholders are involved, including business experts, consumer associations, nongovernmental organizations (NGOs), academics, governmental authorities and, in some cases, innovation laboratories.

The fact that CR tools are developed by expert groups lends them a certain status from the perspective of both practitioners and the public (Terlaak 2007). The fact that CR tools are perceived by the public as signs of legitimacy also gives them significant symbolic value, a value which has facilitated their proliferation and wide acceptance as a form of self-regulation. It would be inappropriate to reinvent the wheel by devising RRI tools without first determining the potential of widely accepted CR tools. To do so is the purpose of this book.

References

CHGE (n.d.) Climate change and biodiversity loss. Center for Health and the Global Environment at the Harvard T.H. Chan School of Public Health. http://www.chgeharvard.org/topic/climate-change-and-biodiversity-loss. Accessed 21 Apr 2015

European Commission (2013) Call for tenders No RTD-B6-PP-00964-2013: study on moni-
    toring the evolution and benefits of responsible research and innovation—tender specifica-
    tions. European Commission, Directorate-General for Research & Innovation, Directorate
    B—European Research Area, B.6—Ethics and gender. https://infoeuropa.eurocid.pt/files/
    database/000056001-000057000/000056403_2.pdf/. Accessed 21 Apr 2015
Forbes (n.d.) The world's billionaires: 2015 Ranking. http://www.forbes.com/billionaires/
    list/#version:static. Accessed 21 Apr 2015
Grunwald A (2014) A new and emerging research community. ProGReSS eNewsletter #2, Spring 2014.
    http://www.progressproject.eu/wp-content/uploads/2013/07/ProGReSS-eNewsletter_Spring2014.
    pdf. Accessed 21 Apr 2015
Ishii-Eiteman M (2013) You can't put the GE genie back in the bottle. Ground Truth, 12 June
    2013. http://www.panna.org/blog/you-cant-put-ge-genie-back-bottle. Accessed 21 Apr 2015
Naam R (2013) Science will save the planet (if we let it). Wired, 17 May 2013. http://www.
    wired.co.uk/magazine/archive/2013/05/ideas-bank/science-will-save-the-planet-if-we-let-it.
    Accessed 21 Apr 2015
NASA (n.d.) Global climate change: Vital signs of the planet. National Aeronautics and Space
    Administration. http://climate.nasa.gov/causes/. Accessed 21 Apr 2015
Porter ME, Kramer MR (2011) Creating shared value. Harvard Business Review, Jan–Feb 2011.
    https://hbr.org/2011/01/the-big-idea-creating-shared-value. Accessed 22 Apr 2015
Schroeder D (2011) Does the pharmaceutical sector have a co-responsibility to secure the human
    right to health? Camb Q Healthc Ethics 20(2):298–308. doi:10.1017/S0963180110000952
Simon S (2012) Childhood leukaemia survival rates improve significantly. American Cancer
    Society, 27 March 2012. http://www.cancer.org/cancer/news/childhood-leukemia-survival-
    rates-improve-significantly. Accessed 21 Apr 2015
Terlaak A (2007) Order without law? The role of certified management standards in shaping socially
    desired firm behaviours. Acad Manag Rev 32(3):968–985. doi:10.5465/AMR.2007.25275685
UNEP (n.d.) Climate change: Introduction. United Nations Environment Programme. http://www
    .unep.org/climatechange/Introduction.aspx. Accessed 21 April 2015
United Nations (1948) Universal Declaration of Human Rights. United Nations. http://www.
    un.org/en/documents/udhr/index.shtml. Accessed 22 Apr 2015
United Nations (n.d.) We can end poverty: Millennium Development Goals and beyond 2015.
    United Nations. http://www.un.org/millenniumgoals/global.shtml. Accessed 22 April 2015

Chapter 2
The Basics of Responsible Research and Innovation

Abstract Responsible research and innovation (RRI) is a newly emerging governance framework, promoted initially by public funders of research. This chapter explains the concept by defining its individual elements (responsibility, research and innovation). Three case studies are given: one from South Africa, where indigenous community involvement provided a significant lead for a health innovation; one from Germany, where end-user involvement in the innovation process led to faster and less contentious market entry; and one from India, where an innovation significantly improved the lives of the poorest girls and women. The concepts of responsiveness, inclusiveness and providing a societal good are illustrated through the case studies, mapped against policy and academic work on RRI and derived from the earlier discussions of responsibility.

Keywords Responsible research and innovation · Responsibility · Responsiveness · Inclusiveness · Societal good

One could reasonably argue that responsible research and innovation (RRI) is a down-to-earth concept for researchers, innovators, citizens and policymakers alike. Unlike other concepts in science and innovation governance, it is expressed in familiar terms: 'responsible', 'research' and 'innovation'. As such the contrast with more technical concepts such as midstream modulation (Fisher et al. 2006), upstream engagement (Rogers-Hayden and Pidgeon 2007) and real-time (Guston and Sarewitz 2002) or constructive technology assessment (Schot 1992) is considerable.

However, if one takes a closer look there are highly interesting and complex discussions hidden within the concept of RRI. The first part of this chapter will therefore explain the basic terms that constitute RRI. The second will ask 'responsibility for what?' and explore why RRI is of interest to the business community. In the third part technical definitions of RRI will be introduced to elucidate what the term means specifically to researchers and innovators. The chapter concludes with a summary and recommended further reading.

© The Author(s) 2016

K. Iatridis and D. Schroeder, *Responsible Research and Innovation in Industry*,
SpringerBriefs in Research and Innovation Governance,
DOI 10.1007/978-3-319-21693-5_2

2.1 Defining Research, Innovation and Responsibility

2.1.1 Research and Innovation

Research is systematic investigation (observation, experiment, critical thinking), which aims to increase knowledge and reach new conclusions. It is a very broad term. For instance, if somebody stands in front of a public library every morning to see how many people enter between 9.30 and 10.00 and then derives some conclusion from this observation (e.g. the library might as well only open at 10.00, because nobody ever comes between 9.30 and 10.00), one could call this research. At the same time, the search for the Higgs boson[1] at the Large Hadron Collider in Switzerland, estimated to have cost around US\$13.25 billion (Knapp 2012), is also research. A research question or target is needed, as well as a systematic investigation and the attempt to derive new conclusions.

Innovation, on the other hand, is a more specific concept and more closely related to business and industry. It can be described as a process of using information and existing phenomena to improve human lives by creating better products, services and technologies that are readily available to markets, governments and society (Stahl et al. 2013). In the context of writing about *responsible* innovation, it has been defined as follows.

> Innovation is an activity or process which may lead to previously unknown designs pertaining either to the physical world (e.g. designs of buildings and infrastructure), the conceptual world (e.g. conceptual frameworks, mathematics, logic, theory, software), the institutional world (social and legal institutions, procedures and organization) or combinations of these, which - when implemented - expand the set of relevant feasible options for action, either physical or cognitive (van den Hoven 2013).

Talking about responsible research and innovation therefore means talking about *applying the concept of responsibility* to deliberate actions performed in the process of increasing knowledge and reaching new conclusions through systematic investigations, or through the development of new products, processes, technologies or services. What does 'responsibility' mean then?

2.1.2 Responsibility

The term 'responsibility' goes back to the Latin *respondere* or *respondum*, which was used in Roman courts to refer to the justification or defence of certain actions and inactions (Schwartländer 1974, p. 1579). Some philosophers broaden the term

---

[1]An elementary particle in physics.

to include responsibility for opinions and the adoption of values.[2] However, we will restrict our discussion here to the responsibility for action or non-action. The term then implies that a

> person had a reason, or reasons, to perform some action, then formed an intention to perform that action (or not perform it) and finally acted (or refrained from acting) on that intention, and did so on the basis of that reason(s) (Miller 2011, p. 138).

To explain what this means in more detail and to distinguish a range of different senses of responsibility, let us look at criticism of the textile industry. We have chosen an example that fits into the theme of corporate responsibility (see Chap. 3) rather than RRI debates, given that no innovators are involved, in order to ensure that the better-known concept of corporate responsibility can help us understand the newer concept of RRI.

In an article on ethical consumerism, a campaigner and author[3] asks what one should call fashion produced in an environmentally friendly manner and without exploitative labour conditions. Green fashion? Fair fashion? Ecofashion? Ethically correct fashion? He continues:

> 'Ethically correct fashion? I am tired of hearing this,' a young fashion designer recently said to me. ... To work with materials free of poison? To pay staff fairly and not let them labour endlessly? To produce clothes one can dispose of without environmental problems? 'Why are we not looking for words for those who do not accept these standards?'... How far have we come if we laboriously have to designate and certify what should be the most obvious thing in the world? (Grimm 2015) (our translation)

The fashion designer may be right, but the label 'ethically correct fashion' would probably be inappropriate, for example, for large parts of the Bangladeshi ready-made garment sector, where the following practices are widespread:

> informal recruitment, and irregular payment, sudden termination, wage discrimination, excessive work, and abusing child labour. Moreover workers suffer various kinds of diseases due to the unhygienic environment and a number of workers are killed in workplace accidents, fires and panic stampedes (Ahamed 2012, p. 3).

In addition, corruption is very widespread in Bangladesh. On a corruption perceptions index by Transparency International (2015), it came 145th out of 175 countries and territories.

The ready-made garment industry contributes a vast proportion of Bangladesh's exports (78 %) and employs 3.6 million people, 2.8 million of whom are women (Ahamed 2012, p. 2). By comparison, the most dominant industrial sector in Germany, automotive engineering, employs 756,000 people (Make it in Germany n.d.). Which levels of responsibility for practices within this industry can be distinguished?

---

[2]For instance, Raz (2001) argues that people are responsible for their own biases and 'other distortions of their cognitive functioning' and therefore are also responsible for failed judgements coloured by bias. He calls this epistemic responsibility.

[3]Grimm, author of *Shopping hilft die Welt verbessern* (2009).

First, we have natural or individual responsibility, which can be ascribed to the person who is most immediately involved with a given practice, for example somebody who recruits child labour locally, accepts bribes or dismisses staff ad hoc.

Second, we have those in a superior institutional role to the individually responsible person; those who are 'responsible for the actions of other persons in virtue of being the person in authority over them' (Miller 2011, p. 139). In smaller companies this will be the chief executive officer, but in larger companies it may be the local or regional managing director. Both are likely to support existing business practices, if they are widespread.

Third, some responsibility for given practices is commonly given to local governments. For instance, when a fire killed 110 people in a Bangladeshi garment factory just outside Dhaka in 2012, a trade union representative in Pakistan noted that it 'is the responsibility of the local governments to make sure that the labour laws are properly implemented in these factories' (Shams and Birkenstock 2012).

Fourth and fifth, the last two levels of responsibility that can be identified in the case of the Bangladeshi garment industry look at those who benefit directly from these harmful practices, in particular powerful international retailers and their consumers. For instance, after the Rana Plaza[4] tragedy, a workers' rights pressure group noted: 'Anybody sourcing in Bangladesh should be aware this could be happening in their supply chain' (Butler 2013). The trade union representative in Pakistan who commented after the 2012 fire said:

> Countries like Bangladesh and Pakistan face tough competition from other markets that provide cheap labour to international companies. They compromise on safety measures to reduce their services cost. It works well for international retailers as they are there to make a profit (Shams and Birkenstock 2012).

Those who profit from practices that foreseeably create harm must carry some responsibility for those practices (Schroeder 2011). This is best argued with reference to the most severe harm of human death. Although philosophers have been debating the intricacies of moral theories for millennia, there is broad agreement that avoidable deaths constitute a harm (Nagel 1979; Feldman 1991) and that foreseeable harm must be avoided by those who have the power to do so, such as international buyers. In virtue theory, '[b]eing able to live to the end of a human life of normal length; not dying prematurely' (Nussbaum 2000, p. 78) is given prime importance, as the first human capability that exerts a moral claim on others. In utilitarian theory, it is taken for granted that '[s]uffering and death … are bad' and that we must 'prevent something bad from happening' (Singer 2009, p. 15). In rights-based theory it is assumed that the 'state of Nature has a law … to govern it;

---

[4]On 24 April 2013, an eight-storey commercial building collapsed in Savar, near Dhaka, the capital of Bangladesh. The building hosted mostly garment factories; the accident killed 1129 people and left many more with serious disabilities (Butler 2013).

… no one ought to harm another in his life' (Locke 1690). Hence, bad things happening in one's supply chain, as the Bangladeshi trade unionist put it, need action by the responsible manager or firm.

It is not only retailers that benefit, but also the end users, in this case the consumers of cheap garments. However, discussions of this type of collective responsibility are more controversial. For instance, it is unclear whether consumers have an obligation to obtain the knowledge required to buy ethically. In this context, some also argue that ethical consumers can never substitute for strong workers' rights locally (Esbenshade 2004). Figure 2.1 summarizes the various levels of responsibility for action or non-action in the Bangladeshi garment industry.

How would these responsibility levels look to researchers and innovators? Let us assume a team of innovators is working in a private laboratory developing skin-care products based on properties found in marine invertebrates. As the research team needs to remove marine invertebrates from their natural habitat to extract

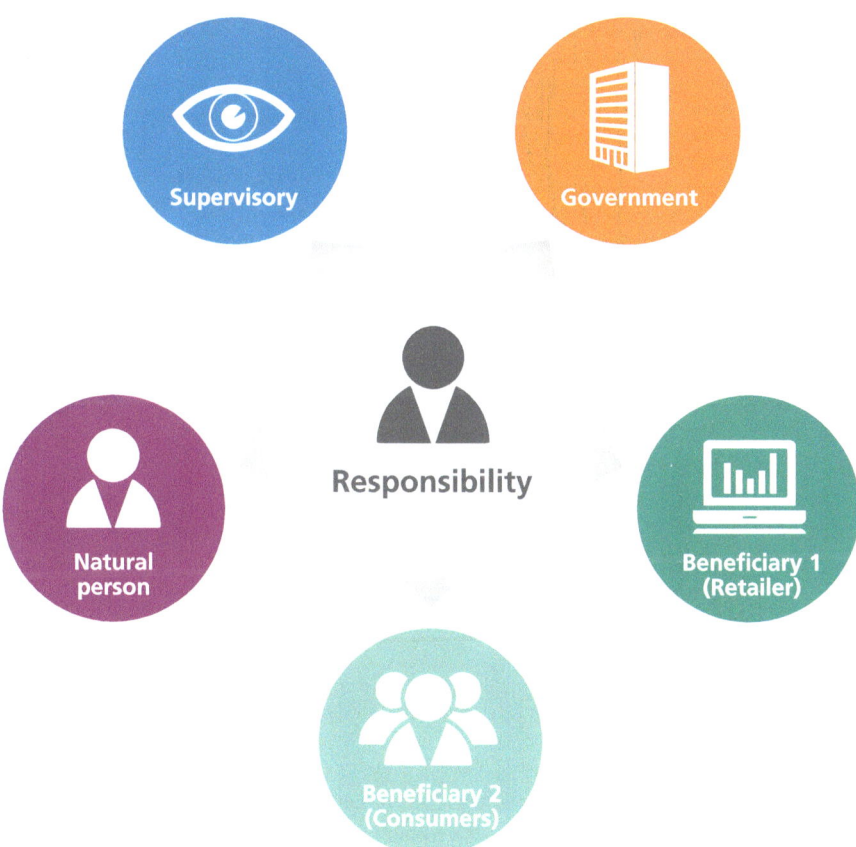

Fig. 2.1 Levels of responsibility

promising biomolecules, the innovation process involves risks for the environment, especially since some of the invertebrates are on the CITES[5] list of endangered species. To match the Bangladeshi example, we now assume that the harmful practice can be identified, in this case that an endangered species has been further depleted, and additionally that the research team is not native to the area where the species was harvested, but works with local collaborators.

Natural or individual responsibility for the resulting environmental harm then lies with the harvesters, who are the local collaborators. Supervisory responsibility lies with the research team that funds and directs the harvesting process. Governmental responsibility is easily identified in this example, as 180 governments have agreed to become parties to CITES (n.d.). It is then their responsibility to adopt domestic legislation to ensure that the convention is implemented. A failure to adopt adequate compliance mechanisms may have led to the overharvesting in this example, for which the local government carries some responsibility. As for the responsibility level of beneficiaries, we again have two groups of beneficiaries in this example: those who will eventually commercialize any skincare products and benefit financially and those who consume the products that were produced while depleting biodiversity.

To understand the concept of responsibility fully, one further distinction is important, namely the difference between legal, contractual and moral responsibility (Werner 2013, p. 41). Within the law, 'legal responsibility' is often considered the informal term for 'legal liability', which means that a person can be expected to do or refrain from doing something, as prescribed by the law. If somebody does not act as the law requires, he or she can be brought to a civil or a criminal court. For instance, unfair dismissal would be decided in a civil court (or, in the UK, for instance, by an employment tribunal[6]). The benchmark against which employee and employer actions would be judged is the labour laws of a given country.

It is also possible that a dispute may arise between employer and employee about the interpretation of a contract. For instance, employees on management contracts might believe they deserve a bonus for 'exceptional performance', as laid down in the contract, but there may be a disagreement on whether performance was exceptional or not. (Most employers therefore try to provide some kind of metric for the assessment of performance.) Such a case would normally be adjudicated between the contracting parties, with the help of human resources advisors or other mediators. One then speaks of contractual responsibilities.[7]

---

[5]CITES (the Convention on International Trade in Endangered Species of Wild Fauna and Flora) is an international agreement with 180 parties. Its aim is to ensure that international trade in specimens of wild animals and plants does not threaten their survival (http://www.cites.org).

[6]Employment tribunals are set up by the UK government specifically in order to resolve disputes between employers and employees over employment rights.

[7]Contract law is a framework for regulating voluntary exchange transactions, such as employment contracts. One may wonder why legal and contractual responsibilities should be separated. However, the distinction is not only routinely made (Werner 2013:41), but helpful in untangling the complexity of obligations companies face (see next section).

Fig. 2.2 Types of responsibility

Finally, moral responsibilities in the work context are the most difficult to define. For instance, while there are clear guidelines on what is ethically *un*acceptable in the work context (e.g. sexual harassment), there are no guidelines on the level of care or empathy one should manifest.[8] Bereavement leave, for instance, is unregulated and is within the decision-making capacity of the employer. Figure 2.2 illustrates the three types of responsibility in a Venn diagram to show that they can overlap. For instance, the bereaved employee might be able to produce a certificate from a sympathetic doctor testifying that she cannot work due to stress. The same case then moves into the labour law arena, in accordance with health and safety laws.

2.2 Responsibility for What? And Why Should This Be of Interest to Industry?

'Responsibility' is a term that invites the question: for what? In this sense it is a term like 'obligation' or 'accountability'. Obligation for what? Accountability for what? And that question can only be answered in context or very broadly. For instance, parents are generally responsible for the safety and wellbeing of their children; CEOs are generally responsible for the profitability and legality of their companies. These are broad answers, which are unlikely to guide practical action. More contextual and therefore specific answers to the question 'responsibility for what?' would be that Employee X is responsible for the compilation of daily production statistics and their comparison with budgeted figures, and that Nurse Y is responsible for the blood sugar measurements on Ward Z.

---

[8]For an excellent argument on moving care and empathy into the centre of ethical thinking, see Slote (2007).

Table 2.1 Specificity of Responsibilities

Responsibility	Specificity	
Contractual	High	Contracts between two or more parties are usually very specific, and it is in the interests of all parties to understand their respective roles and obligations (i.e. their responsibilities), as exploitation of the lack of knowledge of a party can lead to the annulment of a contract.[a] For instance, an employment contract should specify not only remuneration and other benefits for the employee, but also the employer's expectations in terms of performance and responsibilities
Legal	Medium	Given that laws apply to much larger groups than contracts, their specificity is usually lower. In addition, awareness of a law is not immediate, as in a contract, which each party needs to sign. This may also be the case because companies can operate across national borders within different legal frameworks. While a research team that works with marine specimens can be expected to have heard of CITES (the Convention on International Trade in Endangered Species of Wild Fauna and Flora), the Zembrin® case study below shows that a lot of initiative might be required to be aware of new legislation
Moral	Low	Moral responsibilities are most difficult to specify concretely. What are a company's precise moral responsibilities for distant suppliers such as those in the Bangladeshi example? To supervise their activities as one would supervise the activities of one's own employees? To ensure that local laws are applied? And if so, how?

[a]For instance, a party who has been persuaded to agree to a contract without fully understanding his or her role and responsibilities might be able to claim unconscionable dealing, meaning that the better-informed and more powerful party took unfair advantage of the weaker party

Here it is helpful to recall the distinction between types of responsibility, namely contractual, legal and moral. Table 2.1 explains why, generally speaking, contractual responsibilities are most specific, followed by legal responsibilities, followed by moral responsibilities.

To conclude this section:

Responsibility for what? Responsible researchers and innovators should discharge their contractual, legal *and* moral responsibilities.

2.2.1 The Importance of Inclusiveness and Responsiveness

One can expect researchers and innovators to know their contractual responsibilities. However, they are not always aware of their legal and moral responsibilities. For instance, the Convention on Biological Diversity (CBD) (UNEP 1992) includes the legal responsibility on industry (and others) to share benefits with the holders of traditional knowledge should such knowledge be used in the innovation process. This legal responsibility was made more concrete with the Nagoya

Protocol (CBD 2010). The protocol, in turn, required incorporation into domestic (or equivalent) law. In Europe this occurred in 2014 with Regulation (EU) No 511 (European Parliament 2014). Hence, even though benefit sharing with traditional knowledge holders has been required since 1992, more concrete European law was only issued 22 years later.

At the same time, international civil society organizations[9] and the media (Hindu 2012) became increasingly interested in 'biopiracy', or the misappropriation of traditional knowledge in contravention of the CBD. Innovators who used traditional knowledge without sharing benefits were named and shamed in the international media. For instance, the British *Guardian* titled an article about the alleged biopiracy of traditional knowledge in the Kalahari: 'In Africa the Hoodia cactus keeps men alive. Now its secret is "stolen" to make us thin' (Barnett 2001). The article noted:

> [I]t appears that while the drug companies were busy seducing the media, their shareholders and financiers about the wonders of their new drug, they had forgotten to tell the bushmen, whose knowledge they had used and patented. Phytopharm's excuse appears to be that it believed the tribes which used the Hoodia cactus were extinct.

Being aware of one's legal responsibilities *and* discharging them is thus clearly necessary to avoid serious reputational damage. The case study that follows recounts a case of compliance with the CBD that required considerable initiative and responsiveness from the industry partner.

2.2.1.1 Case Study: Zembrin®—Taking the Initiative on Legal Compliance

In 1986, while in Australia, South African doctor and botanist Nigel Gericke came across a reference to the *Sceletium* plant in a book entitled *Narcotic Plants* by William Emboden. He was intrigued to read about this plant, native to his home country, and that it may have an influence on the central nervous system. In 1992, dried plant material was given to him by an ethnobotanist, and he tried various doses on a circle of professional friends, including doctors, psychiatrists and psychologists. In 1995, Dr Gericke engaged South Africa's leading addictionologist, Dr Greg McCarthy, to accompany him on a visit to two communities in Namaqualand[10] where the plant was still in use. The psychiatrist undertook formal structured interviews using the DSM III criteria for addiction to find out whether the plant's use was addictive. Based on the interviews, it was provisionally concluded that the plant was likely to be safe and non-addictive. A year later, in 1996, an active component isolated from the plant was identified and found to be a

---

[9]See, for instance, Argumedo and Pimbert (2006).

[10]An arid region extending through Namibia and South Africa.

potent 5-HT uptake inhibitor, and a patent based on this activity[11] was filed shortly afterwards. It took almost a decade before a plant extract could be produced that was precisely standardized in terms of both the content and relative composition of the key active compounds. At this point, in 2006, a venture capitalist, Hall Investments, invested in further research, forming HG&H Pharmaceuticals.

In the early 1990s, when researching medical innovation based on botanicals in general, Dr Gericke had read about benefit sharing with traditional knowledge holders by the USA start-up Shaman Pharmaceuticals more than a decade before the South African Biodiversity Act was promulgated in 2004. As the founding director of the Traditional Medicines Programme (TRAMED) of the Department of Pharmacology at the University of Cape Town, he included recognition of indigenous intellectual property rights as one of the objectives of TRAMED. He took the initiative and decided that, to avoid raising unrealistic expectations,[12] he would only make contact with traditional knowledge holders to discuss a benefit-sharing agreement when:

- they could grow the correct chemotype of the plant on a commercial scale;
- the pre-clinical and clinical safety looked good; and
- there would be sufficient capital to support ongoing research and development and commercialization.[13]

This point came in 2007, and since Dr Gericke had been researching the requirement for benefit sharing independently, he was able to make contact with a prominent human rights attorney, who in turn identified an established organization that was able to represent the interests of the primary traditional knowledge holders. It took just under half a year from the formal opening of the benefit-sharing negotiations in October 2007 to the conclusion of a benefit-sharing agreement in February 2008. Parties to the agreement were the San peoples of South Africa, through the South African San Council (SASC), and HG&H Pharmaceuticals, the newly founded company developing an antidepressant and antianxiety botanical extract from *Sceletium tortuosum*. The following benefits were agreed:[14]

- Should commercialization be successful, 5 % of all sales of the extract would be paid into a trust fund for the San peoples.
- An additional 1 % of all profits would be paid into the trust fund for the use of a San logo. The company helped register a trademark for the logo selected by the SASC.

---

[11]5-HT, also known as serotonin, is a hormone and neurotransmitter, an imbalance of which is likely to play a role in depression. When its reuptake is inhibited, more serotonin is available within the body.

[12]This is an important point in dealing with benefit sharing (see Wynberg et al. 2009).

[13]Personal communication from Nigel Gericke to Doris Schroeder, 2 April 2015.

[14]Importantly, the San representatives decided that the income from this benefit-sharing agreement was to be shared fifty-fifty with the two communities in Namaqualand that had provided the information on the addiction potential of *Sceletium*.

- During the first three years after the conclusion of the agreement, at least 250,000 South African rand (approximately €19,000) was to be paid per annum up front, before the first sales of Zembrin® were made.

Meanwhile, traditional knowledge holders contributed to the innovation process not just by providing the initial research lead on the plant and its nonaddictive character, but also by identifying a nonintoxicating variety of the plant, which was then cultivated in the commercialization process. In September 2012, a product containing Zembrin®, the brand name of the standardized and characterized commercial extract of *Sceletium tortuosum* developed by HG&H Pharmaceuticals, was launched in the South African market. In March 2013, the first products containing Zembrin® were launched in the US market. In March 2015, HG&H were granted their third US patent on *Sceletium*, and by this time some 26 branded dietary supplement products containing Zembrin® were on the US market. In addition, Zembrin® had been formally approved for sale by Health Canada as a nonprescription health product.

Responsible research and innovation involve proactively seeking information about legal conduct as well as *doing the right thing*, whether there is a compliance mechanism or not. This requires innovators to be responsive to stakeholders from other areas; in the above case from the law, from civil society organizations and from indigenous peoples' organizations. It also requires an approach to innovation that recognizes the ideal of inclusiveness. The innovation process in the above example was started on the basis of traditional knowledge.

In addition, the innovator included the traditional knowledge holders in the innovation process as well as in the distribution of proceeds derived from the innovation. As a result of this approach, and in comparison with the *Hoodia* case cited above, faster and less contentious market entry was possible, as adverse media publicity was avoided and available knowledge used effectively.

What if an innovator faces no legal obligations? Can an inclusive and responsive approach also be successful without legal responsibilities? A second case will illustrate that faster and less contentious market entry does not have to be prompted by legal requirements.

2.2.1.2 Case Study: Ambiact—Using Stakeholder Engagement for Faster, More Efficient Market Entry

One of the grand challenges of affluent nations are ageing societies (PACITA n.d.). A German start-up company (Oldntec n.d.) was successful in developing an innovative smart metre for social alarm systems, which allows elderly people to stay in their own homes for longer without compromising safety.[15] Instead of relying on the active summoning of help in the case of accidents, the metre monitors whether a standard appliance (e.g. kettle, toaster) is unused for an untypically long time,

---

[15]All information for this case study was taken from Responsible-Industry (n.d.).

usually 24 h. The product, which can be placed between any appliance and the power socket, is connected to a social alarm operator via a care phone and will generate an emergency call if the chosen appliance is not used. Users feel an increased quality of life since the monitoring does not invade their privacy (as cameras, for example, would), nor does it require the daily handling of care phones, which elderly people with incipient memory problems are likely to find onerous.

From the initial idea to a legally approved product took only three years, from 2011 to 2014. The owners of the start-up company believe that 'the cooperation of all stakeholders during the development process was a key factor in the successful development of the ambiact' and that:

1. early engagement of all stakeholders saves costs during development since necessary changes in the product – in terms of acceptability – can be made early in the development process and occasionally in later stages,
2. the possibility for timely feedback on problems or suggestions by stakeholders increases their willingness to participate in the development free of charge and with enthusiasm, and
3. science education is not only helpful for subsequent easier adoption of the new product in the care process, but is also a simple and effective marketing tool for the new product (Responsible-Industry n.d.).

Stakeholder engagement included interviews with end-users as well as social alarm operators at the start of the innovation process to determine the priorities of each group. Further interviews were undertaken before field trials to determine the daily habits of end-users. Field trials helped refine the prototype, and discussions with research participants revealed some interesting requirements. For instance, to the great surprise of the innovators, colour turned out to be the most serious problem with the device for the end-users. The initial black made the device too noticeable to visitors, drawing attention to the possible frailty of the householder, while the white, proposed later, tended to show the dirt. A light grey device proved acceptable to all. The company owners concluded:

Overall, the ambiact was developed from an initial idea to the final product in only three years with comparatively little cost by using the 'work force' of volunteer end-users (Responsible-Industry n.d.).

To conclude this section: of the three types of responsibilities we distinguished (contractual, legal and moral), we have taken for granted that researchers and innovators know their contractual responsibilities. They enter the contract and hence have to be aware of it. We have also seen how initiative, responsiveness and inclusiveness can help fulfil legal responsibilities while potentially leading to faster and less contentious market entry. Plus we have shown that such faster and less contentious market entry can be achieved through inclusiveness in the innovation process, independent of legal requirements. This leaves moral responsibilities to discuss.

2.2.2 The Importance of Societal Good

In moral philosophy, one usually distinguishes 'do no harm' from 'do good' responsibilities. Philosophers might call this the difference between negative and positive duties (Berlin 1969) or the difference between perfect and imperfect duties (Kant 1998). A typical 'do no harm' duty would be 'do not kill', while a typical 'do good' duty would be 'help those in need'. Not to kill somebody, ignoring special cases like self-defence, usually requires no effort and not much thought. One just needs to refrain from doing it. To help those in need, by contrast, requires a lot of thought and often considerable effort. Whom do we help? Only those in our sphere of familiarity (family and friends?) or distant strangers? If the latter, why? For both, *what* can we do to help and where do our responsibilities end; i.e. when have we done enough?

Such debates are usually far removed from the industrial sphere and especially from research and innovation. Science governance frameworks preceding RRI focused almost exclusively on 'do no harm'. For instance, the background and foundation of technology assessment was formed by 'negative experiences with the development and use of new technologies', especially 'unintended consequences and catastrophic accidents' (Dusseldorp 2013) (our translation). It is therefore unsurprising that technology assessment's main focus is new and emerging technologies such as nanotechnology, synthetic biology and, still, genetic modification. Likewise, corporate responsibility usually focuses on 'do no harm', while 'do good' is separated out as 'corporate philanthropy'. Figure 2.3 illustrates this graphically.

If responsible research and innovation is based on contractual, legal *and* moral responsibility, as the term 'responsible' requires, the 'do good' element of moral responsibility cannot be ignored. What does this mean for innovators?

2.2.2.1 Case Study: The Indian Sanitary Pad Revolutionary—Doing Good Through Innovation

It is easy to find straightforward, but also striking, examples of innovators who have made a huge difference to people's lives and done enormous good. For instance, in 2014, an Indian man from a poor family made headlines around the world as he 'revolutionised menstrual health for rural women in developing countries by inventing a simple machine they can use to make cheap sanitary pads' (Venema 2014). Years earlier, as a newlywed, he had been appalled by his wife's use of dirty cloths during menstruation and set out to invent cheap alternatives that poor women, like his wife and widowed mother, could afford. (In 2011, only 12 % of Indian women could afford disposable sanitary pads, while, at the same time, around '70 % of all reproductive diseases in India are caused by poor menstrual hygiene—it can also affect maternal mortality' (Venema 2014)).

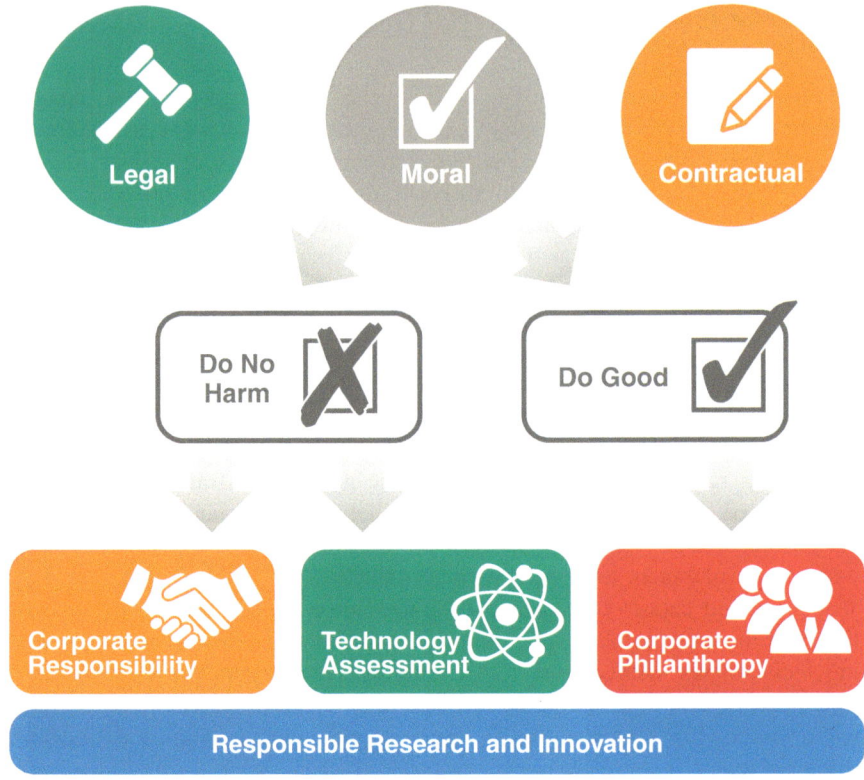

Fig. 2.3 Moral responsibility in context

It took Arunachalam Muruganantham almost five years to create a cheap method of producing sanitary pads, but in the end the social entrepreneur invented a low-cost machine that could be operated with little training and used with locally available materials. Regarded as having breached the taboos surrounding menstruation in India, he was believed to be possessed by evil spirits, and thus was deserted by his wife and mother and required to leave his village (Venema 2014). Subsequently, his wife and mother returned and he was given an innovation prize by the Indian president.

> He was once asked whether receiving the award from the Indian president was the happiest moment of his life. He said no – his proudest moment came after he installed a machine in a remote village in Uttarakhand, in the foothills of the Himalayas, where for many generations nobody had earned enough to allow children to go to school. A year later, he received a call from a woman in the village to say that her daughter had started school. 'Where Nehru failed,' he says, 'one machine succeeded' (Venema 2014).[16]

This may also be the reason why *Time* magazine put him on its list of the most influential 100 people in 2014 (Gupta 2014). Meanwhile, the innovation has also reached half a million African girls and women through Afripads (AFRIpads n.d.).

---

[16]In rural India, girls are likely to drop out of school with the onset of menstruation.

Fig. 2.4 The bottom of the pyramid

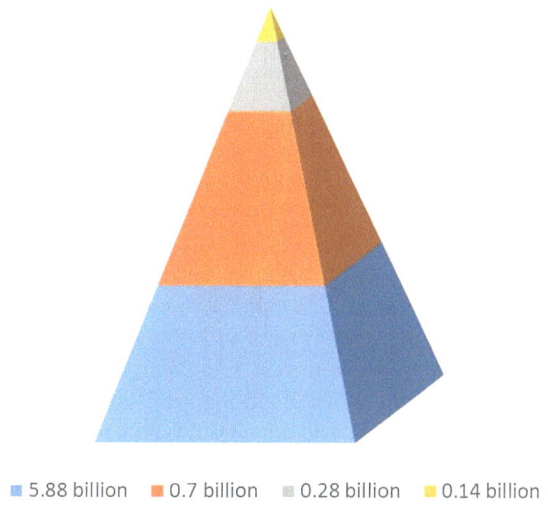

■ 5.88 billion   ■ 0.7 billion   ■ 0.28 billion   ■ 0.14 billion

Of course, not all innovators can revolutionize the life of the poor. At the same time, using the term 'responsible' in RRI means that one cannot ignore this aspect of moral responsibility. And it is not ignored in the business community. More recent efforts among industry to bring innovation closer to society talk about 'shared value' and link it with business success. As a senior Harvard Professor has written in the *Harvard Business Review*:

> The solution [to society's disenchantment with industry] lies in the principle of shared value, which involves creating economic value in a way that also *creates value for society by addressing its needs and challenges*. Businesses must reconnect company success with social progress. Shared value is not social responsibility, philanthropy, or even sustainability, but a new way to achieve economic success (Porter and Kramer 2011).

References to reputational gain are common when the business literature writes about philanthropy or even, as here, about new, more socially embedded ways of doing business. But is that enough to persuade innovators to focus more strongly on societal needs?

2.2.3 The Fortune at the Bottom of the Pyramid

Economists differentiate between four tiers of consumers, represented in a pyramid (Fig. 2.4).

The four tiers of the pyramid each have the same annual income, namely 25 % of US$70 trillion in 2012 (Alexander 2012).[17] In 1998, C.K. Prahalad first wrote about combining corporate success with eradicating poverty and claimed:

> This is a time for MNCs [multi-national corporations] to look at globalization strategies through a new lens of inclusive capitalism. For companies with the resources and persistence to compete at the bottom of the world economic pyramid, the prospective rewards include growth, profits, and incalculable contributions to humankind (Prahalad and Hart 2002).

In the fifth edition of a book now recognized as ground-breaking, *The Fortune at the Bottom of the Pyramid*,[18] Prahalad presents case studies of enterprises—in fields ranging from finance to energy to high-tech solutions—that help people escape poverty while providing a commercial profit. Most importantly, he explains that 'the rate of diffusion among the Bottom of the Pyramid around the world has shown how willing and capable the poor are to accept and benefit from advanced technology' (Prahalad 2014). An example Prahalad gives is the mobile phone; and indeed, a study commissioned by the World Bank and undertaken in Kenya showed that, in 2012:

- Over 60 % of the respondents among the Kenyan BoP [bottom of pyramid] own a mobile phone …
- 1 in 4 Kenyan BoP mobile phone owners use Internet on their mobile phone …
- a quarter of … study respondents stated that they had earned money through the use of their mobile phone … because they were more 'reachable' …
- no difference [was found] in mobile phone activities between men and women …
- health and education Information [was] most desired (Crandall et al. 2012).

Speaking within the framework of moral responsibilities, one could therefore say, as Prahalad does, that 'doing good' and 'doing well' (making a profit) are not mutually exclusive (Prahalad 2014). This is pertinent, of course, in responding to businesses that ask why RRI should be relevant to them. Table 2.2 summarizes the twelve principles of innovation for BoP markets.[19]

As this is a short book for industry, it does not focus on social and community innovations for the bottom of the pyramid, except for the example above of Indian sanitary pads. This aspect has been dealt with excellently in an article by Mario Pansera and Richard Owen (2014).

Now that we have analysed the term 'responsibility' in detail with all its implications, we will ask: does it relate to academic and policy discussions of RRI?

---

[17]Income distribution data through personal communication from Thomas Pogge based on 2011 data from Branko Milanovic, World Bank. In 5 % steps, the shares are as follows: 0.130, 0.199, 0.248, 0.297, 0.349, 0.413, 0.493, 0.600, 0.741, 0.920, 1.167, 1.515, 1.976, 2.587, 3.396, 4.514, 6.678, 11.520, 19.487 and 42.768 %; world population at the time 7 billion.

[18]It has been argued that 'base of the pyramid' would be a more politically correct term. However, since the majority of authors use the term 'bottom of the pyramid', we have chosen to do so as well, for consistency with the existing literature.

[19]All information in the table from Prahalad (2014:52–71).

Table 2.2   The twelve principles of innovation for BoP markets

Principles	Examples
Price performance	Mobile phones used to cost Indians US$1000. Then Reliance introduced its Monsoon Hungama for an initial payment of $10 and monthly payments of $9.25. Within ten days, one million applications were received
Innovation hybrids	Watered-down versions of ToP (top of pyramid) market products are often not a solution to BoP needs. Hindustan Lever limited (HLL), a subsidiary of Unilever, undertook in-depth research to ensure that iodized salt kept its iodine content in the harsh conditions of storage and transport in India. As a result a product was developed that could help the 70 million Indian children, and others, who suffer from iodine deficiency disorder. HLL is now entering markets in Ghana, Ivory Coast and Kenya with this innovation
Scale of operations	The basis for the BoP market success is volume, not price. Hence, operations must be supported by organizations that have good geographical reach
Sustainable development	When serving a potential market of five billion people, the solutions currently enjoyed by the ToP are not sustainable. However, pursuing such solutions, rather than being a hindrance, might allow all markets to achieve more innovative and sustainable products and services
Identifying functionality	Businesses innovating for the BoP should first understand that functionalities in BoP markets might differ. For instance, India has 5.5 million amputees, with up to 30,000 added each year (through accidents, polio and war). Yet a prosthetic limb must serve in a very different environment from that to which traditional prosthetic limbs are suited. For instance, customarily squatting on the floor, never wearing shoes in temples and having minimal time for fittings must all be taken into account
Process innovation	How things may be done more innovatively while building on local infrastructure can make a big difference in service success. The Aravind Eye Care System, says founder Dr Venkataswamy, aimed to achieve the quality and consistency of the McDonald's chain for eye operations, e.g. cataract surgery. It offers 'eye camps' in local villages, organizes patient transport to Aravind hospitals, and practises a clear and innovative work process for surgeons and post-operative care and counselling
Deskilling of work	Educated labour is difficult to obtain in BoP settings. Voxiva, a Peruvian start-up, developed a device that can be given to rural health care workers, helping them to diagnose a large range of diseases by, for instance, providing photos of the various stages of smallpox. The device also allows rural health workers to contact authorities in Lima

(continued)

Table 2.2 (continued)

Principles	Examples
Education of consumers	As the large majority of BoP consumers have very limited education, giving the poor access to information can create win-win situations. HLL, the Unilever subsidiary mentioned above, used ultraviolet dirt and bacteria detectors to demonstrate to children in village schools that their hands were still dirty after washing in contaminated water. Simply washing one's hands with soap before eating can reduce the incidence of death from diarrhoea by 50 % (more than two million Indian children die from stomach-related illnesses, including diarrhoea, every year). This initiative also served to increase HLL's volume of soap sales
Designing for hostile infrastructure	Products and services must work in hostile conditions, if they are to be viable for BoP markets: for example, actual mains voltages in Indian villages range from 90 to 350 volts
Interfaces	First-time users of new technologies are likely to require special support. For instance, the Mexican retailer Elektra introduced fingerprint recognition at their ATMs so that they would not have to remember nine-digit codes
Distribution	Product and process innovations are important, but in BoP markets, distribution solutions are equally important to access the customer. For instance, Avon cosmetics is extremely successful with its direct sales technique in Brazil, where Avon representatives become mini-suppliers
Challenge conventional wisdom	All the examples in this table challenge conventional wisdom and encourage innovators to embrace new paradigms

2.3 Introduction of RRI Frameworks and Approaches

The European Commission is continuing to support a programme which links research and innovation to societal concerns and interests. The 'Science with and for Society' (SwafS) programme has produced one of the most influential RRI definitions in Europe:

> RRI is an inclusive approach to research and innovation (R&I), to ensure that societal actors work together during the whole research and innovation process. It aims to better align both the process and outcomes of R&I, with the values, needs and expectations of European society. In general terms, RRI implies anticipating and assessing potential implications and societal expectations with regard to research and innovation (European Commission n.d.).

In addition to the above definition, the SwafS unit has developed five measureable action lines to help implementation. These are:

- engage society more broadly in its research and innovation activities;
- increase access to scientific results;

- ensure gender equality, in both the research process and research content;
- take into account the ethical dimension; and
- promote formal and informal science education (European Commission n.d.).

Another definition, which was developed within the European Commission by Rene von Schomberg, notes that RRI is a

> transparent, interactive process by which societal actors and innovators become mutually responsive to each other with a view on the (ethical) acceptability, sustainability and societal desirability of the innovation process and its marketable products (in order to allow a proper embedding of scientific and technological advances in our society) (von Schomberg 2013).

The most widely cited academic work on RRI points to the necessity of common efforts and respect for future generations. 'Responsible innovation is a collective commitment of care for the future through responsive stewardship of science and innovation in the present' (Owen et al. 2013). In implementing responsive stewardship, four RRI dimensions are necessary: anticipation, reflection, deliberation and responsiveness (see Fig. 2.5).

All three definitions, as well as the five action lines, have a *process* and an *outcome* dimension.

Process—Undertaking research responsibly is a collective undertaking. 'Scientists and innovators play an important role, but responsible innovation must be a holistic approach across the innovation ecosystem' (Owen et al. 2013).

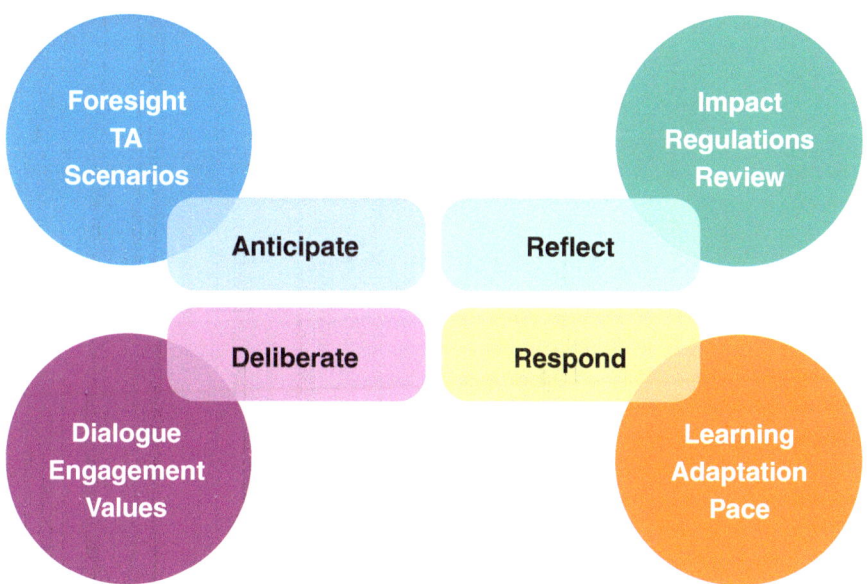

Fig. 2.5 Responsive stewardship

Table 2.3 RRI definitions and responsibilities

RRI statements	Responsiveness	Inclusiveness	Societal good
For the European Commission RRI is an inclusive approach to research and innovation (R&I), to ensure that societal actors work together during the whole R&I process. It aims to better align both the process and outcomes of R&I with the values, needs and expectations of European society. In general terms, RRI implies anticipating and assessing potential implications and societal expectations with regard to research and innovation, and working to: • engage society more broadly in its research and innovation activities • increase access to scientific results • ensure gender equality, in both the research process and research content • take into account the ethical dimension • promote formal and informal science education	Societal actors: • Work together during the whole R&I process • Anticipate and assess potential implications and societal expectations • Engage society more broadly	An inclusive approach Ensuring gender equality	Better aligning both the process and outcomes of R&I with the values, needs and expectations of European society
René von Schomberg RRI is a transparent, interactive process by which societal actors and innovators become mutually responsive with a view to the (ethical) acceptability, sustainability and societal desirability of the innovation process and its marketable products (in order to allow a proper embedding of scientific and technological advances in our society)	Societal actors and innovators become mutually responsive		The societal desirability of the innovation process and its marketable products
Richard Owen et al. Responsible innovation is a collective commitment of care for the future through responsive stewardship of science and innovation in the present. Scientists and innovators play an important role, but responsible innovation must be a holistic approach across the innovation ecosystem. The first and foremost task for responsible innovation is then to ask what futures we collectively want science and innovation to bring about	Responsive stewardship	A holistic approach across the innovation ecosystem	The futures we collectively want science and innovation to bring about

This includes universities, research funders, civil society organizations and, importantly, the general public, and requires a co-responsibility for research and innovation.

Outcomes—A framework for RRI goes beyond risk assessment, foresight and control of what we do *not* want research and innovation to do (Owen et al. 2013). It includes the purposes of what we do want research and innovation to do, that is, the envisaged positive outcomes. 'The first and foremost task for responsible innovation is then to ask what futures do we collectively want science and innovation to bring about … ?'(Owen et al. 2013) Governance should then do a lot more than close down ethically unacceptable and unsustainable research. Instead, it should be about defining and realizing 'areas of public value for innovation' (Wildson et al. 2005) and 'benefits to humanity' (Ozolina et al. 2012). A recent report on RRI commissioned by the European Commission notes that:

> The need to gear the innovation process to societal needs is reflected in many high-level policy, strategy and programming documents, such as the objective of the EU 2020 strategy to create smart growth or the Horizon 2020 programme that defines tackling societal challenges as one of the main priorities (van den Hoven et al. 2013).

To see how the definitions above map onto the responsibilities for responsiveness, inclusiveness and a drive towards societal goods, refer to Table 2.3.

2.4 Conclusion

Responsible conduct is made up of three elements. It complies with contractual obligations, with legal obligations and with moral obligations. This statement applies in research and innovation as well as in all other business areas. Traditionally, corporate responsibility was seen most pressingly as 'do no harm': no harm to workers, no harm to the environment and no harm to the local community through, for instance, bribery. 'Do good' was regarded as a bonus point, the preserve of corporate philanthropy.

As the RRI frameworks introduced in the last section show, responsible research and innovation draws no significant distinction between *do no harm* and *do good*. For instance, the SwafS definition of RRI asks innovators to align their undertakings 'with the values, *needs* and expectations' of society (Horizon 2020 n.d.). In particular, the term 'needs' has an implicit reference to doing good. Citizen needs usually go beyond not being harmed. Basic needs are generally understood as those required for survival: food, shelter, clothing and access to health care to prevent unnecessary morbidity and mortality (Sen 1992). Some of the case studies introduced in this chapter have a direct bearing on citizen needs. The Indian sanitary pads have a direct impact on health and well-being; Zembrin® and the ambiact also address the health and wellbeing market. Hence, some innovators do indeed align their actions with the needs of society.

Table 2.4 Distinction between ethical acceptability, sustainability and societal desirability

RRI element	Definition with reference to innovation	Identifiable through
Ethical acceptability	Innovation which respects fundamental values without discrimination	Codes of conduct, ethics guidelines and sustained public engagement efforts
Sustainability	Innovation 'that meets the needs of the present without compromising the ability of future generations to meet their own needs' (DEPweb 2001)[a]	Environmental protection and health and safety as, for instance, in ISO standards
Societal desirability	Innovation which can benefit all without discrimination	Addressing citizen needs or grand challenges of humankind

[a]Here we are borrowing Brundtland's definition of 'sustainable development' and applying it to 'sustainable innovation'

According to René von Schomberg, RRI consists of three elements: ethical acceptability, sustainability and societal desirability. While one could argue that a nonpolluted environment is societally desirable and ethically acceptable, and therefore that 'do no harm' has precedence in all three areas, a more refined understanding would separate the three terms. Table 2.4 sets out one possible way of defining the three areas as separate.

In this interpretation of von Schomberg's definition of RRI, one could align 'do no harm' roughly with ethical acceptability and sustainability and 'do good' with societal desirability. Importantly, though, both are present. In comparison with corporate responsibility, therefore, RRI is potentially both broader in scope, in that it demands a link to citizen needs and societal desirability, and smaller in scope, because it deals only with research and innovation rather than the entire business cycle. Figure 2.6 illustrates this in the form of a simplified diagram.[20]

Finally, the Owen et al. definition of RRI hopes that responsible innovation will ensure the kind of future that citizens collectively want science and innovation to bring about, thereby encompassing both the 'do no harm' and 'do good' precepts of responsibility.

Can existing corporate responsibility tools cope with this new shift towards responsible research and innovation? The next chapter will first outline the basics of corporate responsibility before moving towards the answer to this question.

---

[20]For more information on corporate responsibility, see Chap. 3.

Fig. 2.6 RRI and corporate responsibility—relevant parts of the business cycle

2.5 Recommended Reading

- Grunwald (2013) Handbuch Technikethik Verlag JB Metzler, Stuttgart
- Owen et al. (2013) Responsible innovation: Managing the responsible emergence of science and innovation in society. John Wiley, London
- Pansera M, Owen R (2014) Framing resources-constrained innovation at the 'bottom of the pyramid': Insights from an ethnographic case study in rural Bangladesh. Technological Forecasting and Social Change 92:300–311
- Ozolina Z, Mitcham C, Schroeder D, Mordini E, Crowley J, McCarthy P, von Schomberg R (2012) Ethical and regulatory challenges to science and research policy at the global level. Directorate-General for Research and Innovation, European Commission, Brussels

- Pavie X, Scholten V, Carthy D (2014) Responsible innovation: From concept to practice. World Scientific Publishing, Singapore
- Prahalad CK (2014) The fortune at the bottom of the pyramid: Eradicating poverty through profits. Pearson Education, Upper Saddle River, NJ
- van den Hoven J, Jacob K, Nielsen L, Roure F, Rudze L, Stilgoe J (2013) Options for strengthening responsible research and innovation. Directorate-General for Research and Innovation, European Commission, Brussels
- van den Hoven J, Doorn N (eds) (2014) Responsible innovation 1: Innovative solutions for global issues. Springer, Dordrecht

References

AFRIpads (n.d.) AFRIpads [blog]. http://afripads.com/blog/. Accessed 5 May 2015
Ahamed F (2012) Improving social compliance in Bangladesh's readymade garments industry. Labour Manage Dev 13. http://www.nla.gov.au/openpublish/index.php/lmd/article/viewFile/2269/3148. Accessed 3 May 2015
Alexander R (2012) Where are you on the global pay scale? BBC News Magazine, 29 Mar 2012. http://www.bbc.com/news/magazine-17512040. Accessed 5 May 2015
Argumedo A, Pimbert M (2006) Protecting indigenous knowledge against biopiracy in the Andes. International Institute for Environment and Development, London
Barnett A (2001) In Africa the Hoodia cactus keeps men alive. Now its secret is 'stolen' to make us thin. The Observer, 17 Jun 2001. http://www.theguardian.com/world/2001/jun/17/internationaleducationnews.businessofresearch. Accessed 5 May 2015
Berlin I (1969) Four essays on liberty. Oxford University Press, Oxford
Butler S (2013) Bangladeshi factory deaths spark action among high-street clothing chains. The Observer, 23 Jun 2013. http://www.theguardian.com/world/2013/jun/23/rana-plaza-factory-disaster-bangladesh-primark. Accessed 3 May 2015
CBD (2010) Nagoya protocol on access to genetic resources and the fair and equitable sharing of benefits arising from their utilization to the convention on biological diversity. http://www.cbd.int/abs/. Accessed 3 May 2015
CITES (n.d.) List of contracting parties. http://www.cites.org/eng/disc/parties/chronolo.php. Accessed 3 May 2015
Crandall A, Otieno A, Mutuku L, Colaço J, Grosskurth J, Otieno P (2012) Mobile phone usage at the Kenyan base of the pyramid. Final report, iHub Research, Research Solutions Africa. https://blogs.worldbank.org/ic4d/files/ic4d/mobile_phone_usage_kenyan_base_pyramid.pdf. Accessed 5 May 2015
DEPweb (2001) What is sustainable development? The World Bank Group. http://www.worldbank.org/depweb/english/sd.html. Accessed 5 May 2015
Dusseldorp M (2013) Technikfolgenabschätzung. In: Grunwald A (ed) Handbuch Technikethik. Verlag JB Metzler, Stuttgart, pp 194–399
Esbenshade J (2004) Monitoring sweatshops: workers, consumers, and the global apparel industry. Temple University Press, Philadelphia
European Commission (n.d.) Science with and for society. Horizon 2020: The EU Framework Programme for Research and Innovation. http://ec.europa.eu/programmes/horizon2020/en/h2020-section/science-and-society. Accessed 5 May 2015
European Parliament (2014) Regulation (EU) No 511/2014 of the European Parliament and of the Council of 16 Apr 2014 on compliance measures for users from the Nagoya protocol on access to genetic resources and the fair and equitable sharing of benefits arising from their utilization in the union. Off J Eur Union. L 150/59. http://eur-lex.europa.eu/legal-content/EN/TXT/HTML/?uri=CELEX:32014R0511&from=EN. Accessed 3 May 2015

Feldman F (1991) Some puzzles about the evil of death. Philos Rev 100:205–227

Fisher E, Mitcham C, Mahajan R (2006) Midstream modulation of technology: governance from within. Bull Sci Technol Soc 26:485–496. doi:10.1177/0270467606295402

Grimm F (2009) Shopping hilft die Welt verbessern (Shopping helps improve the world). Mosaik Verlag, München

Grimm F (2015) Vom Irrsinn, der als Normalität durchgeht. Schrot und Korn, January 2015:58. http://schrotundkorn.de/lebenumwelt/lesen/vom-irrsinn-der-als-normalitaet-durchgeht.html. Accessed 3 May 2015

Gupta R (2014) The 100 most influential people: Arunachalam Muruganantham. Time, 23 Apr 2014. http://time.com/70861/arunachalam-muruganantham-2014-time-100/. Accessed 5 May 2015

Guston D, Sarewitz D (2002) Real-time technology assessment. Technol Soc 24:93–109

Hindu (2012) Stress on media's role in fighting bio-piracy. The Hindu, 6 Mar 2012. http://www.thehindu.com/todays-paper/tp-national/tp-kerala/stress-on-medias-role-in-fighting-biopiracy/article2965633.ece. Accessed 5 May 2015

Kant I (1998) Groundwork of the metaphysics of morals. Cambridge University Press, Cambridge

Knapp A (2012) How much does it cost to find a Higgs boson? Forbes, 5 Jul 2012. http://www.forbes.com/sites/alexknapp/2012/07/05/how-much-does-it-cost-to-find-a-higgs-boson/. Accessed 10 Jan 2015

Locke J (1690) The second treatise of civil government http://libertyonline.hypermall.com/Locke/second/second-frame.html. Accessed 3 May 2015

Make it in Germany (n.d.) Automotive engineering. Federal Ministry for Economic Affairs and Energy, Federal Ministry of Labout and Sodial Affairs, Bundesagentur für Arbeit. http://www.make-it-in-germany.com/en/for-qualified-professionals/working/industry-profiles/automotive-engineering. Accessed 3 May 2015

Miller S (2011) Collective responsibility and global poverty. In: Boylan M (ed) The morality and global justice reader. Westview Press, Philadelphia, pp 135–151

Nagel T (1979) Death. In: Nagel T (ed) Mortal questions. Cambridge University Press, Cambridge, pp 1–10

Nussbaum M (2000) Women and human development: the capabilities approach. Cambridge University Press, Cambridge

Oldntec (n.d.) Welcome to oldntec. http://www.oldntec.de/en/. Accessed 5 May 2015

Owen R, Stilgoe J, Macnaghten P, Gorman M, Fisher E, Guston D (2013) A framework for responsible innovation. In: Owen R, Bessant J, Heintz M (eds) Responsible innovation: managing the responsible emergence of science and innovation in society. John Wiley, London, pp 27–50

Ozolina Z, Mitcham C, Schroeder D, Mordini E, Crowley J, McCarthy P, von Schomberg R (2012) Ethical and regulatory challenges to science and research policy at the global level. Directorate-General for Research and Innovation, European Commission, Brussels

PACITA (n.d.) European stakeholder involvement in ageing society. Parliaments and Civil Society in Technology Assessment. http://www.pacitaproject.eu/ageing-society/. Accessed 5 May 2015

Pansera M, Owen R (2014) Framing resources-constrained innovation at the 'bottom of the pyramid': insights from an ethnographic case study in rural Bangladesh. Technol Forecast Soc Chang 92:300–311

Porter M, Kramer M (2011), Creating shared value. Harvard Business Review, January-February 2011. http://hbr.org/2011/01/the-big-idea-creating-shared-value. Accessed 5 May 2015

Prahalad CK, Hart SL (2002) The fortune at the bottom of the pyramid. Strategy + Business 26. http://www.stuartlhart.com/sites/stuartlhart.com/files/Prahalad_Hart_2001_SB.pdf. Accessed 5 May 2015

Prahalad CK (2014) The fortune at the bottom of the pyramid: eradicating poverty through profits. Pearson Education, Upper Saddle River

Raz J (2001) Engaging reason. Oxford University Press, Oxford

Responsible-Industry (n.d.) Results of the bottom-up call for case study descriptions. http://www.responsible-industry.eu/activities/bu-casestudies-results. Accessed 5 May 2015

Rogers-Hayden F, Pidgeon N (2007) Moving engagement 'upstream'? Nanotechnologies and the Royal Society and Royal Academy of Engineering's inquiry. Publ Underst Sci 16(3):345–364. doi:10.1177/0963662506076141

Schot J (1992) Constructive technology assessment and technology dynamics: the case of technology dynamics. Sci Technol Hum Values 17:36–56

Schroeder D (2011) Does the pharmaceutical sector have a co-responsibility to secure the human right to health? Camb Q Healthc Ethics 20(2):298–308

Schwartländer J (1974) Verantwortung. In: Krings H, Baumgartner HB, Wild C (eds) Handbuch philosophischer Grundbegriffe. Koesel Verlag, München, pp 1577–1588

Sen A (1992) Inequality re-examined. Oxford University Press, Oxford

Shams S, Birkenstock G (2012) Bangladesh factory tragedy: Who's responsible? Deutsche Welle, 26 November 2012. http://www.dw.de/bangladesh-factory-tragedy-whos-responsible/a-16406189. Accessed 3 May 2015

Singer P (2009) The life you can save. Text Publishing Company, Melbourne

Slote M (2007) The ethics of care and empathy. Routledge, London

Stahl B, Eden G, Jirotka M (2013) Responsible research and innovation in information and communication technology: identifying and engaging with the ethical implications of ICTs. In: Owen R, Bessant J, Heintz M (eds) Responsible innovation. John Wiley, London, pp 199–218

Transparency International (2015) Corruption by country/territory: Bangladesh. http://www.transparency.org/country#BGD. Accessed 3 May 2015

UNEP (1992) Convention on Biological Diversity: Text and annexes. United Nations Environment Programme, Chatelaine, Switzerland. http://www.cbd.int/. Accessed 3 May 2015

van den Hoven J (2013) Value sensitive design and responsible innovation. In: Owen R, Bessant J, Heintz M (eds) responsible innovation. John Wiley, London, pp 75–83

van den Hoven J, Jacob K, Nielsen L, Roure F, Rudze L, Stilgoe J (2013) Options for strengthening responsible research and innovation. Directorate-General for Research and Innovation, European Commission, Brussels

Venema V (2014) The Indian sanitary pad revolutionary. BBC News Magazine, 4 Mar 2014. http://www.bbc.com/news/magazine-26260978. Accessed 5 May 2015

von Schomberg R (2013) A vision of responsible research and innovation. In: Owen R, Bessant J, Heintz M (eds) Responsible innovation: managing the responsible emergence of science and innovation in society. John Wiley, London, pp 51–74

Werner M (2013) Verantwortung. In: Grunwald A (ed) Handbuch Technikethik. Verlag JB Metzler, Stuttgart, pp 38–43

Wilsdon J, Wynne B, Stilgoe J (2005) The public value of science. Demos, London

Wynberg R, Schroeder D, Chennells R (2009) Towards best practice for community consent and benefit sharing. In: Wynberg R, Schroeder D, Chennells R (eds) Indigenous peoples, consent and benefit sharing. Springer, Berlin, pp 335–350

Chapter 3
The Basics of Corporate Responsibility

Abstract A concise account of corporate responsibility (CR) is provided to facilitate a later comparison of CR and responsible research and innovation. The chapter clarifies the concept by discussing different types of corporate responsibilities and explaining their links with similar concepts such as corporate sustainability, corporate accountability, corporate citizenship and corporate social performance. One focus is on how to develop a CR strategy and the main topics that firms need to pay attention to when doing so. Another focus is the benefits stemming from engagement in CR activities.

Keywords Corporate responsibility · Business case · Business strategy · Benefits

In a world where the three richest companies in 2014 achieved revenues greater than the gross domestic product (GDP) of the poorest 60 countries (IMF 2014; PWC 2014), the mounting societal pressures on business to accept social and environmental responsibilities come as no surprise. These pressures have greatly contributed to the expansion of the concept of *corporate responsibility* (CR). Without doubt, we now live in a corporate world in which the role of business is not limited to profit maximization, but needs to take ethical and social imperatives into account too. This chapter clarifies the concept of CR, offers a roadmap for developing a CR strategy and highlights key aspects and benefits of CR.

3.1 Defining Corporate Responsibility

The context of CR has changed over the years. From Friedman's definition that 'there is one, and only one, social responsibility for business—to use its resources and engage in activities designed to increase its profits so long as it remains engaged in open and free competition, without deception or fraud' (Friedman 1962, p. 133)—we have passed to Drucker's approach that the topic involves active

© The Author(s) 2016   31
K. Iatridis and D. Schroeder, *Responsible Research and Innovation in Industry*,
SpringerBriefs in Research and Innovation Governance,
DOI 10.1007/978-3-319-21693-5_3

commitment and bringing a positive change in one's community, one's society, and one's country (Drucker 1993). More recently, the European Union defined CR as 'the responsibility of enterprises for their impacts on society' (European Commission 2011, p. 6) and maintained that a successful CR approach is based on stakeholder engagement strategies that address social, environmental, ethical, human rights and consumer concerns in business strategies and operations.

As may be deduced, recent CR conceptualizations are explicitly anthropocentric, in that they place human welfare as top priority, in contrast with past notions of CR, which assigned priority fully to business and whatever benefited business. Attention has shifted from shareholders' interests to those of other stakeholders too, such as employees, suppliers, nongovernmental organizations, government, customers and local communities.

Despite this shift, it would be wrong to say that modern perceptions of CR downplay the importance of satisfying shareholders' interests and ensuring a healthy profit. These undoubtedly remain important aspects of business, since no company can survive without financial capital; it would not be able to pay its shareholders, employees and suppliers. What is more, without financial capital a company cannot invest in new products or processes, which would seriously affect its competitiveness. On the contrary, contemporary notions of CR recognize the importance of economic issues, but go beyond them. The rationale underlying modern views is that businesses operate not in isolation, but within society as a whole. As such, they need to address the social and environmental impacts of their business operations and manage them in an ethical way that does not harm stakeholders.

Today, a key challenge for businesses is that stakeholders' interests can vary significantly. For instance, an employee's main concern might be a satisfactory salary and safe working conditions, a local community's need might be a clean and non-polluted environment, and a customer's priority might be fair pricing and safe products. To address this diversity of stakeholder needs effectively, companies have to integrate social and environmental issues into their business models and not treat those issues as separate from their business strategy. This necessitates finding the right balance of economic, social and environmental imperatives.

Hence, apart from acknowledging an economic responsibility to satisfy shareholders' interests, contemporary notions of CR highlight social and environmental responsibilities (Fig. 3.1). Examples of issues pertaining to social responsibility are human rights, labour, anti-corruption, the protection of information, health and safety, and non-discrimination. Equally, environmental responsibilities include topics like pollution abatement, the protection of biodiversity, energy conservation, climate change and product stewardship.

At this point we need to make a distinction between CR and corporate philanthropy. The latter refers to charitable aid and donations and represents a company's way of giving something back to its local or international community. These donations can be either financial or nonfinancial. Financial donations can take the form of fundraising or may be a percentage of a firm's profits given back to communities. Nonfinancial donations are product donations, such as computers,

medicines and food vouchers, and contributions to employee volunteerism. Both types account for a major corporate give-back to communities that in 2013, in the United States alone, accounted for US$18.7 billion (Gose and Frostenson 2014). Notwithstanding the value of this corporate contribution to society, it would be wrong to conflate corporate philanthropy with CR, simply because many companies that engage in charity and donations might not be taking their economic, social and environmental responsibilities into account at the same time. Put differently, such companies might be using corporate philanthropy to enhance their corporate image and distract from their business operations. Thus, corporate philanthropy is a narrower topic than CR, and a charitable company is not necessarily a responsible one.

3.2 Similar Concepts to CR

The CR movement has paved the way for the development of other related concepts, such as corporate sustainability, corporate accountability, corporate citizenship and corporate social performance. Box 3.1 presents a brief description of these topics. All these developments describe one thing; companies are adopting strategies that do not only maximize profits, but also target the social and environmental impacts of their operations.

> **Box 3.1 Concepts related to CR**
>
> **Corporate sustainability** mostly focuses on eco-efficiency and the critical use of natural resources, so that future generations too have access to these resources. Eco-efficiency is a combination of economic and ecological efficiency and implies the production of more goods and services with less energy and lower consumption of natural resources, resulting in less waste and pollution.
> **Corporate accountability** goes beyond voluntary engagement in enhancing corporate social and environmental performance and implies enforceability, meaning that businesses should be held accountable for the social and environmental impacts of their operations.
> **Corporate citizenship** implies that just as active citizens take care of issues in their local communities, companies too need to be active 'citizens' and address the social and environmental impacts of their business operations.
> **Corporate social performance** pays particular attention to the outcomes of businesses' attempts to address social issues and stakeholders' concerns and focuses on the tools companies use to report on their performance.

3.3 Developing a CR Strategy

In the past, a company could enjoy a healthy competitive advantage by innovating and simultaneously enjoy the trust of the public. Due to the tendency of modern societies to put their faith in expert systems (Giddens 2007), coming up with something new was enough for businesses to claim specialized knowledge and effectiveness in managing their operations. As Habermas maintains (1987), in modern societies authority relationships are based not on caste, class or age but on effectiveness. Members who have the knowledge (in our case businesses) claim that they are effective in managing a situation and those who do not have any knowledge (stakeholders) trust them.

The key difference nowadays is that this relationship has broken down and innovation alone is not enough to make stakeholders trust businesses. Companies are increasingly called upon to address the negative impacts of their operations and come up with responsible, innovative solutions that are embedded in business strategy. Firms that embody stakeholder engagement in their business strategy and are skilful at innovating responsibly gain a competitive advantage over their rivals, while companies that lack these skills gradually see their survival being threatened.

An impediment to a corporate response to this new business environment is the difficulty companies experience in decoding the CR agenda and breaking it down into a series of topics. Without doubt this is a crucial step for implementing CR; yet the large number of existing definitions and the variety of concepts and approaches have perplexed businesspeople seeking to apply a responsible business model. In order for a company to address its corporate responsibilities effectively it needs to focus on the following:

- **Business impacts**: A business needs to identify the social and environmental impacts of its operations and assess their significance.
- **Tools/policies**: There is a pool of tools available nowadays for mitigating those impacts, ranging from management standards to global initiatives. A business needs to choose the appropriate tool for mitigating the impacts stemming from business operations. For instance, if a company belongs to the manufacturing sector and its operations entail environmental impacts, then it needs to adopt an environmental management standard to reorganize its business activities and minimize those impacts;
- **Stakeholders' concerns**: As mentioned above, stakeholders have different views and concerns. However, not all of them are of equal importance. A business must identify the most important stakeholders, prioritize their concerns and then set up an action plan for stakeholder engagement. The latter is of major importance as it enables companies to understand stakeholders' interests and develop a strategy that satisfactorily addresses these concerns.

3.4 Main CR Topics

Corporate responsibility is an area in which more than legal compliance is at stake. Businesses are rendered socially responsible when they invest in the human factor and in the environment, in the interests of all stakeholders. Of the vast array of CR topics, six stand out as the most important and provide a useful map for companies that want to address their corporate responsibilities. These are:

- **Quality**: Ensure that products and services are safe for use and do not include any hazardous substances that might degrade the environment or endanger human health.
- **Environmental protection**: Adopt policies that promote energy saving, waste management and recycling, pollution abatement, the minimum consumption of raw materials and water, noise control, and the protection of local biodiversity.
- **Health and safety**: Develop policies that ensure that all health and safety issues are fully addressed, risks are minimized and legal requirements are not violated.
- **Labour**: Respect employees' rights to negotiation, avoid child employment and discrimination, ensure humane working hours and abide by labour laws and international labour standards.
- **Anticorruption**: Guarantee transparency in all business transactions, avoid bribery and disclose information on any private or political contributions and donations made or received.

- **Data protection**: Protect private or sensitive information and data from any unethical use and manage issues related to privacy, software and intellectual property.

3.5 The Business Case for CR

According to the Oxford English Dictionary (2015), a business case is 'a justification for a proposed project or undertaking on the basis of its expected commercial benefit'. One might say: 'We have to put a business case forward to the managers as to why this investment is necessary.' Equally, the term 'the business case for CR' has been used to explain to the corporate world why it makes sense to engage in CR activities. The argument is that a socially responsible profile entails significant advantages for both the company and society.

Numerous studies have researched the link between corporate financial and social performance, and evidence suggests that the application of socially responsible practices and good relationships with stakeholders enhance corporate reputation (Dacin and Brown 1997), improve market value (McWilliams and Siegel 2000), boost a firm's attractiveness to employees (Marin and Ruiz 2007), reduce costs and risks to the firm (Carroll and Shabana 2010) and enhance the firm's operational efficiency (Corporate Watch 2006). Also, engagement in CR activities contributes to social justice, assists in correcting social and environmental problems related to business activities, promotes a dialogue and partnerships among diverse stakeholder groups and safeguards compliance with legal requirements in a variety of areas, such as working conditions, environmental protection, and health and safety. The challenge for a business is to develop a business strategy that balances the aforementioned benefits (Fig. 3.2), as an imbalance in favour of business benefits might harm society and vice versa (Coombs and Holladay 2012).

Well-known companies support the CR movement, as they realize the imperative in doing so. Colgate-Palmolive (www.colgate.com), for instance, has realized the potential of CR to improve both the social and financial performance of the corporation, and this is why it promotes more responsible and innovative business conduct. The company's approach consists of three main pillars, namely people, performance and planet.

'People' refers to initiatives and corporate policies developed to engage stakeholders in business operations. One noteworthy initiative in this field is the nine consumer innovation centres the company set up to ascertain consumers' preferences and generate products according to their needs. Equally, examples of stakeholder-related policies include the supplier code of conduct and specific policies on antibribery, product safety research, ingredient safety and no deforestation.

In the 'performance' domain, the firm provides financial information and data on the actions it undertakes to manage its corporate responsibilities effectively and produce products that have a positive social and environmental impact. A key aspect of the firm's strategy on this front is product sustainability. To achieve this,

Fig. 3.2 CR benefits

the company is working towards improving ingredient biodegradability and the recycling of packaging. Furthermore, Colgate-Palmolive has created its own product sustainability scorecard to track progress on improving the sustainability profile of new products, and at the same time set up its own sustainability excellence award, which is given to company divisions that provide evidence of integrating Colgate's sustainability strategy into all aspects of their operations.

Finally, under 'planet', the company includes actions undertaken to combat the climate change phenomenon, and its planet-related commitments to address energy consumption, water use, waste management and greenhouse gas emissions.

All this work on CR has been widely recognized, as the company was included in Ethisphere's 2014 list of the World's Most Ethical companies, received the award for the Best Global Green Brand for 2013, and was granted the United States Environmental Protection Agency Energy Star for 2014.

3.6 Conclusion

Slowly but surely, businesses are abandoning traditional shareholder value approaches and adopt more responsible and sustainable conduct. Contemporary societal and economic challenges have played a major role in this shift. These developments challenge established business beliefs related to companies' long-term survival and competitive advantage. While in the past innovation was perceived as one of the core, if not the key, means of achieving a sustainable competitive advantage, today companies also need to address stakeholder concerns, initiate stakeholder engagement activities and take into account the social and environmental impacts of their operations.

A successful corporate response to the contemporary business environment necessitates the integration of CR into business operations, the adoption of new strategies and management systems and the production of new products that serve societal needs.

One crucial question at this point is how responsibility gains practical relevance and becomes embedded in corporate structures and processes. This is achieved through the use of tools developed to decode the corporate responsibility agenda and provide guidance to firms on how to engage with CR practices. Such tools are discussed in detail in the next chapter.

References

Carroll A, Shabana K (2010) The business case for corporate social responsibility: a review of concepts, research and practice. Int J Manag Rev 12(1):85–105

Coombs WT, Holladay S (2012) Managing corporate social responsibility: a communication approach. Willey-Blackwell, Oxford

Corporate Watch (2006) What's wrong with corporate social responsibility? Corporate Watch, Oxford

Dacin PA, Brown TJ (1997) The company and the product: Corporate associations and consumer product responses. J Mark 61(1):68

Drucker PF (1993) Post-capitalist society. Butterworth-Heinemann, Oxford

European Commission (2011) Communication from the commission to the European parliament, the council, the European economic and social committee and the committee of the region: a renewed EU strategy 2011–14 for corporate social responsibility. European Commission, Brussels

Friedman M (1962) Capitalism and freedom. University of Chicago Press, Chicago

Giddens A (2007) The consequences of modernity. Polity Press, Cambridge

Gose B, Frostenson S (2014) Corporate profits surge but cash donations creep up only 3 %. Chron Philanthr, 13 July 2014

Habermas J (1987) The theory of communicative action volume 2: lifeworld and system—a critique of functionalist reason. Polity, Cambridge

IMF (2014) World economic outlook database. http://www.imf.org/external/pubs/ft/weo/2014/01/weodata/index.aspx. Accessed 3 February 2015

Marin L, Ruiz S (2007) I need you too! Corporate identity attractiveness for consumers and the role of social responsibility. J Bus Ethics 71(3):245–260

McWilliams A, Siegel D (2000) Corporate social responsibility and financial performance: correlation or misspecification? Strateg Manag J 21(5):603

Oxford English Dictionary (2015) Business case. http://www.oed.com/view/Entry/334243?rskey=7JiXX5&result=1&isAdvanced=false#eid. Accessed 3 February 2015

PWC (2014) Global top 100 companies by market capitalisation. IPO Centre, London

Chapter 4
An Overview of Corporate Responsibility Tools and Their Relationship with Responsible Research and Innovation

Abstract Corporate responsibility (CR) tools have an excellent potential to facilitate the implementation of responsible research and innovation (RRI) in industry. This chapter provides a detailed account of existing CR tools. It develops selection criteria for choosing the tools with the greatest potential to assist the implementation of RRI in industry. Background information on these tools is given, as well as data on the tools' requirements. In addition, assessment criteria are developed to illustrate the tools' suitability in setting up a framework for action that could form a basis for RRI implementation.

Keywords Responsible research and innovation · Business tools · Management standards · Global initiatives · Principles · Codes of ethics

Responsible research and innovation (RRI) implies practical execution, yet there is a dearth of tools dedicated to assisting businesses in implementing RRI principles. What limited attempts have been made to develop suitable instruments have mostly focused on nanotechnology (e.g. Malsch et al. 2012). This represents a significant shortcoming in RRI implementation. We believe that existing tools developed to facilitate the integration of CR into business strategy could significantly improve the situation and contribute towards making RRI a practical reality for firms. More precisely, these CR tools are robust enough to control business conduct effectively and could pave the way for the development of a common framework for RRI, as outlined by the European Commission in its report on options for strengthening responsible research and innovation (European Commission 2013).

4.1 Rationale

There is a plethora of CR tools available, including standards, principles, codes of conduct and initiatives, but not all of them are useful in assisting RRI implementation. To compile a list of the tools with the greatest potential for doing

© The Author(s) 2016   39
K. Iatridis and D. Schroeder, *Responsible Research and Innovation in Industry*,
SpringerBriefs in Research and Innovation Governance,
DOI 10.1007/978-3-319-21693-5_4

so, we consulted three main sources of information, namely the European Commission's core set of internationally acknowledged CR initiatives and principles (European Commission 2011); the relevant CR literature (e.g. Beschorner and Müller 2007; Boiral 2011; Bondy et al. 2008; Cragg 2005; Delmas and Montiel 2008; Jamali 2010; King and Toffel 2009; Locke et al. 2013; Nadvi 2008; Runhaar and Lafferty 2009; Russo 2009; Toffel et al. 2013; Vogel 2010; Leipziger 2010; Heras-Saizarbitoria and Boiral 2013); and international organizations (e.g. the International Organization for Standardization [ISO], the International Labour Organization [ILO] and the Organisation for Economic Co-operation and Development [OECD]). We decided not to include codes of conduct in our list, as:

- they usually focus on a particular firm or industry and do not necessarily take into account the interests of external stakeholders—a crucial issue, as there cannot be responsible innovation without stakeholder involvement;
- their credibility is limited, as companies adhering to codes of conduct are usually subject to internal scrutiny only and are not accountable to a broader constituency (Leipziger 2010); and
- the topics addressed by most codes of conduct are included in international standards, as many codes of conduct turn into international standards once they are tested and matured enough (Leipziger 2010).

Our final choice of the CR tools that could promote RRI implementation was based on the following criteria (Fig. 4.1). The tools had to:

- be international in scope, so that they would provide a common framework for action beyond national regulation;
- endorse a systematic way of dealing with ethical, social and environmental issues;

Fig. 4.1 Criteria for selecting CR tools related to RRI

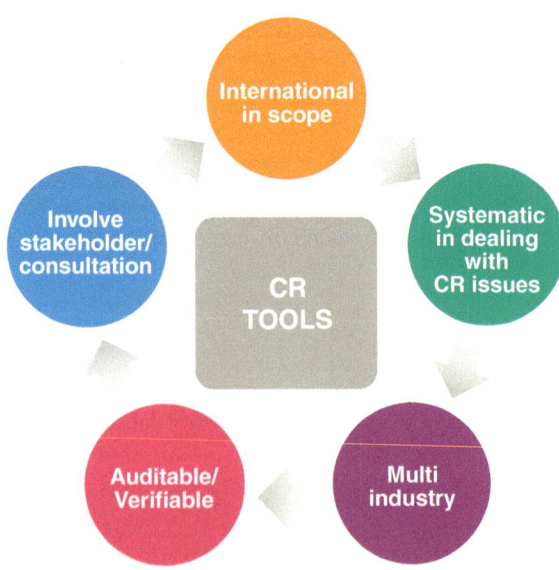

Standards — ISO9001, ISO14001, EMAS, ISO50001, OHSAS18001, SA8000, ISO/IEC27001, ISO26000

Global Initiatives — GRI, Global Compact, The OECD Guidelines for MNE, UN Guiding Principles on Business and Human Rights, ILO MNE Declaration

Principles — Business Principles for Countering Bribery, Caux Round Table Principles, CERES Roadmap for Sustainability, ETI Base Code, Business Social Compliance Initiative

Fig. 4.2 Selected CR tools

- be applicable to a variety of industries irrespective of sector or size;
- be auditable or verifiable through a clear and systematic procedure;
- be the outcome of a wide consultation process, taking into account the interests of both internal and external stakeholders.

4.2 Tools Selected

The CR tools we chose (Fig. 4.2) represent the strongest choices of firms when it comes to managing their corporate responsibilities. Of these, the most popular is ISO9001,[1] with more than 1,100,000 certified firms, followed by ISO14001, with more than 300,000 certifications worldwide (ISO 2013). Other popular choices are the ISO27001 standard, with more than 22,000 certifications (ISO 2013), and the UN Global Compact, with 12,817 signatories (UNGC 2015), while 11,801 companies worldwide have published and registered their CR reports at CorporateRegister.com, a website created and audited by AccountAbility, the drafting entity of the AA1000 series standards (CorporateRegister.com 2015).

The CR tools chosen have several common features. They are compatible, applicable to every organization, regardless of sector or size, and voluntary. The underlying principle of most of the selected CR tools is continuous improvement based on different versions of the Deming cycle, also known as the Plan-Do-Check-Act cycle (Kolesar 1994). This specifies a management strategy for organizations in order to enhance their performance continuously with respect to the issue addressed by the tool, for instance health and safety in the workplace, environmental protection or stakeholder engagement. According to the cycle, companies must initially analyse their current position and set objectives, targets and indicators, and then make plans to achieve them (plan); after that, they must

---

[1]The standards mentioned are described in detail below.

implement those plans (do); next, they have to collect data on their performance against the set objectives and targets (check); and finally, they should apply the necessary corrective actions to overcome any problems (act).

ISO standards, as well as the SA8000 and OHSAS18001 certification standards, have fully encompassed this cycle, whereas other CR tools have adopted variations of it. For example, the AA1000 series uses the Plan-Prepare-Implement-Review and Improve cycle (AccountAbility 2008, 2011), the Global Reporting Initiative (GRI) has adopted the Prepare, Connect, Define, Monitor and Report cycle (GRI 2013), and the Global Compact follows the Commit, Assess, Define, Implement, Measure and Communicate cycle (UNGC 2010). In reality, these all share the same logic of voluntary, continuous improvement that characterizes most CR tools.

Another interesting aspect of the tools selected for our analysis is that they all address stakeholder concerns. The stakeholders most frequently referred to are customers, suppliers, employees and the general public, while other well-known corporate stakeholders, such as shareholders and competitors, are mentioned less often. Additionally, these CR tools require companies to conform to relevant laws and operate in an accountable and truthful manner. In this context, there are recurring provisions for releasing information, either in the form of corporate reports or in documents presented during external audits.

4.2.1 Standards

Most standards are certifiable by third parties, known as certification bodies, and require companies to design, develop and apply a management system. Such a system is intended to demonstrate the ability of businesses to operate their activities in a way that satisfies both societal demands and the applicable regulatory requirements. More specifically, standards require the identification of all processes performed during a firm's operation and the development of documentation for the application of the management system. Depending on the standard, it may, for example, be a quality management system or an environmental management system.

Moreover, these standards entail the allocation of a person or a team to be responsible for the management and monitoring of the system. In most cases, the system's documentation is organized at three levels: strategic, operational and functional. The strategic level defines the management policy with respect to the issue addressed by the standard, and sketches the main principles of its implementation. The operational level involves a number of procedures that are needed to describe the functionality of the processes. A 'procedure' is a documented description of the way certain tasks have to be performed so that the policy and objectives or targets are realized. Hence it is vital that these procedures are clear, consistent with the scheduled activities and frequently reviewed, so that they accurately depit business operations. Also, it is essential that they are comprehensive and provide accurate as well as reliable information. In this context, they describe:

- *the way* each task will be executed;
- *the person* responsible for its execution;
- *the means* needed for the task's execution;
- *the location* of each task;
- *the time* needed for the execution of each task; and
- *the documents* created for the implementation of these actions.

The functional level includes documents that contain either information with which the firm executes certain tasks, or information for products or services that the organization provides. Examples of such documents are: working instructions (where required); job descriptions; internal policies; product specifications; and external documents of the management system including those relating to the normative, legislative and regulatory regime.

The nature and extent of the management system's documentation depends on the size and complexity of the organization and can be in either paper or electronic form, so that documents are easy to access and understand. The CR standards selected for our analysis are described in Tables 4.1, 4.2, 4.3, 4.4, 4.5, 4.6, 4.7, 4.8 and 4.9.

Table 4.1 ISO9001

Summary
ISO9001 focuses on aspects of quality management and provides detailed guidance for companies and organizations interested in continuously enhancing quality aspects of their operations.
Drafting entity
International Organization for Standardization (ISO).
Released
2008 (3rd rev.)—an updated version will be published by the end of 2015.
RRI fit
The standard can facilitate the implementation of RRI in several ways. It improves accountability and transparency through internal and external feedback mechanisms. It also ensures the prevention of consumer abuses in the marketplace by requiring the company not to practise price gouging, make misleading advertising claims or sell ineffective, unreliable or unsafe products. Furthermore, ISO9001 requires the use of safe equipment that does not threaten employees' integrity at all stages of the company's operations (Alic and Rusjan 2010; Frederick 2006).
Core elements
Quality management system; customer focus; leadership; involvement of people; process approach; systems approach to management; continual improvement; factual approach to decision-making; mutually beneficial supplier relationships.
Audit
One annual external audit and provision for internal audits; as a rule of thumb, evidence of at least two internal audits is required during external audit by certification bodies.
Usage to date
1,129,446 certified companies worldwide (ISO 2013).

Table 4.2 ISO14001 and eco-management and audit scheme (EMAS)

Summary	
ISO14001 and EMAS deal with environmental management. The standards establish criteria for the evaluation of the environmental performance of the organization and focus on actions to be taken to minimize the harmful effects this may cause to the environment.	
Drafting entity	
International Organization for Standardization (ISO); European Commission.	
Released	
ISO140001: 2004 (1st rev.)—an updated version will be published by the end of 2015; EMAS: 2009 (2nd rev.)—currently under survey to assess whether a 2015 revision is needed.	
RRI fit	
The use of these two standards contributes towards protecting the rights of local communities through pollution control measures. In addition, they can help companies achieve greater eco-efficiency, greener products (sometimes) and more transparency for and acceptance by external stakeholders (Zwetsloot 2003). The standards require the development of a specific methodology for identifying the environmental aspects of a company's activities and evaluating the environmental impacts stemming from its operations or products. Among the topics that are taken into consideration are air pollution, noise control and waste management (European Commission 2014a; ISO 2009).	
EMAS, especially, aims to provide stakeholders with even more reliable and credible information about the firm's environmental performance than ISO14001, as it requires companies to take additional steps to ensure the continuous improvement of firms' environmental performance. First, EMAS requires the publication of an environmental statement. In it, information on the organization's efforts and achievements is included. More precisely, the environmental statement entails: an assessment of all the significant direct and indirect environmental issues; a summary of year-by-year figures on emissions, waste generation, consumption of raw materials, energy and water, as well as noise; and a presentation of the organization's environmental policy, programmes and management system. Second, the organization's performance on environmental management is checked by environmental verifiers. Third, EMAS requires more active employee involvement in its implementation and monitoring as part of the standard's stakeholder engagement activities (European Commission 2014a).	
Core elements	
Environmental management system; environmental communication; emergency preparedness and response; identification of legal and other normative requirements; identification of environmental aspects and environmental impact assessment; waste management (toxic and nontoxic, solid and liquid); air pollution management; noise control; monitoring and measurement of environmental aspects; environmental statement (EMAS only).	
Audit	
One annual external audit and provision for internal audits; as a rule of thumb, evidence of at least two internal audits is required. Regarding the environmental statement required by EMAS, prior to its publication it must be validated by an accredited verifier, who should not have a previous relationship with the company or organization (European Commission 2014a).	
Usage to date	
301,647 ISO14001-certified organizations (ISO 2013); 13,872 EMAS-registered companies (European Commission 2014b).	

Table 4.3 ISO50001

Summary
ISO50001 deals with energy management; the standard assists organizations in applying the necessary policies for efficiently managing the energy they consume.

Drafting entity
International Organization for Standardization (ISO).

Released
2011.

RRI fit
The standard aims to facilitate the use of energy management best practices and support good energy management policies, so that a reduction in greenhouse gas emissions is achieved. By doing so, the organizations using this standard can conserve resources and contribute to tackling climate change (ISO 2011). This standard is particularly important in contributing towards a sustainable economy, where limited natural resources are available. What is more, the standard can significantly assist organizations in meeting the three major goals defined by the United Nations for 2030 regarding energy use, i.e. ensure universal access to modern energy services; reduce global energy intensity by 40 %; and increase renewable energy use globally to 30 % (ISO 2012).

Core elements
Energy management system; energy policy; energy planning; energy audit; energy performance indicators.

Audit
One annual external audit and provision for internal audits; as a rule of thumb, evidence of at least two internal audits is required.

Usage to date
4,826 certified companies (ISO 2013).

Table 4.4 OHSAS18001 (will change to ISO45001 in 2016)

Summary
This standard aims to: (a) reduce the risks associated with health and safety at work by identifying the health and safety aspects of an organization's activities; (b) minimize the risk of accidents; and (c) diminish any legal violations.

Drafting entity
British Standards Institution (BSI).

Released
2007 (1st rev.).

RRI fit
The implementation of the standard ensures safer and healthier workplaces and the avoidance of negligent practices that may threaten employees' safety. OHSAS18001 can be adopted by any organization wishing to formalize risk management related to health and safety in the working environment for employees, customers and the general public (SGS 2009).

Core elements
Health and safety management system; hazard identification, risk assessment and determining controls; legal and other requirements; objectives and occupational health and safety programme; communication, participation and consultation; operational control; emergency preparedness and response; performance measuring, monitoring and improvement.

(continued)

Table 4.4 (continued)

Audit
One annual external audit and provision for internal audits; as a rule of thumb, evidence of at least two internal audits is required.

Usage to date
92,302 certified companies (OHSAS Project Group 2011).

Table 4.5 SA8000

Summary
SA8000 is based on International Labour Organization (ILO) standards, the Universal Declaration of Human Rights and the UN Convention on the Rights of the Child. It sets out requirements related to workers' rights and workplace conditions.

Drafting entity
Social accountability international (SAI).

Released
2014 (4th rev.).

RRI fit
The standard contributes to achieving greater transparency for companies and their suppliers. It opposes any discrimination based on: race, caste, national origin, religion, disability, gender, sexual orientation, union membership, political affiliation or age. It requires companies to pay their employees, in accordance with relevant laws, wages that ensure the covering of basic needs and provision of a discretionary income. In this way, the adoption of the standard advances the implementation of socially responsible practices. At the moment, it is widely accepted as a standard that secures an ethical environment for employees and its popularity is steadily increasing (Leipziger 2010).

Core elements
Social management system; child labour; forced labour; health and safety; freedom of association and right to collective bargaining; discrimination; working hours; remuneration.

Audit
Semi-annual surveillance audits are conducted; at least one external audit in the first three years must be scheduled.

Usage to date
3,388 certified companies (SAI 2014).

Table 4.6 ISO/IEC27001

Summary
ISO/IEC 27001 is an international standard that focuses on information security management and sets procedures for: (a) the effective management of confidential and sensitive information; and (b) the application of information security controls.

Drafting entity
International Organization for Standardization (ISO), International Electrotechnical Commission (IEC).

Released
2013 (1st rev.).

(continued)

Table 4.6 (continued)

RRI fit

The adoption of ISO27001 reassures the market and stakeholders at large about how a company deals with information. In particular, the standard assists organizations in maintaining the confidentiality, integrity and availability of that information. The latter relates not only to an organization's own assets, but also to information regarding stakeholders including present and past employees and customers. In addition, the adoption of the standard enables organizations to protect private or sensitive information from any unethical use and deal effectively with issues related to privacy, software and data intellectual property issues.

Core elements

Information security management system; security policy; organizational security; asset classification; identification of information security risks; analysis of risks; personnel security; physical and environmental security; communications and operations management; access control; system development and maintenance; business continuity management.

Audit

One annual external audit and provision for internal audits; as a rule of thumb, evidence of at least two internal audits is required.

Usage to date

22,293 certified companies (ISO 2013).

Table 4.7 ISO26000

Summary

This is a non-certifiable standard, meaning it provides guidance on how to apply CR practices, rather than requirements. The intention is to allow space for adopters of the standard to develop their own CR activities according to their own needs, unrestricted by specific requirements which might limit the effectiveness of the CR programme.

Drafting entity

International Organization for Standardization (ISO).

Released

2010 (currently under review).

RRI fit

The use of ISO26000 would foster the implementation of RRI through seven principles of CR (see core elements below) (ISO 2010). The rationale behind the standard is that companies interested in adopting it will prioritize the seven principles of CR according to their own needs and build business models in the spirit of continuous improvement. Interestingly, this standard covers very similar issues to the GRI reporting guidelines (discussed below). In particular, ISO26000 assists companies in organizing their activities in such a way that they can be measured and subsequently presented in the organization's report (GRI 2013).

Core elements

Accountability; transparency; ethical behaviour; respect for stakeholders' interests; respect for the rule of law; respect for international norms of behaviour; respect for human rights.

(continued)

Table 4.7 (continued)

Audit
Although certification bodies offer training programmes for auditors of ISO26000, companies that adopt the standard are not necessarily required to undergo an audit review.

Usage to date
64 countries[a] have adopted ISO26000 as a national CR standard (ISO 2013).

[a]These are: Algeria, Argentina, Austria, Belgium, Bolivia, Bosnia and Herzegovina, Brazil, Bulgaria, Chile, Costa Rica, Cote D'Ivoire, Croatia, Czech Republic, Denmark, Ecuador, Estonia, Finland, France, Germany, Ghana, Guatemala, Honduras, Hungary, Ireland, Italy, Japan, Kazakhstan, Kenya, Korea (Republic of), Kyrgyz Republic, Lebanon, Lithuania, Malawi, Malta, Mauritius, Mexico, Moldova (Republic of), Mongolia, Morocco, Netherlands, Norway, Oman, Panama, Peru, Portugal, Romania, Saint Lucia, Saudi Arabia, Serbia, Slovak Republic, South Africa, Spain, Sweden, Switzerland, Tanzania, Thailand, Trinidad and Tobago, Tunisia, Turkey, Uganda, United Kingdom, Uruguay, USA, Zimbabwe

Table 4.8 AA1000AS

Summary
AA1000 Assurance Standard (AA1000AS) is a reporting standard which aims to assure the credibility and quality of sustainability reports. It is mainly used by third parties offering sustainability assurance services and it is compatible with the framework provided by ISAE 3000, the financial accounting body standard for verifying information on non-financial matters.

Drafting entity
Institute of Social and Ethical Accountability (ISEA), also known as AccountAbility.

Released
2008 (2nd rev.).

RRI fit
AA1000AS highlights the need for accountable information on companies' sustainability performance and adopts an approach that is based on both qualitative and quantitative data. The latter puts emphasis on quantifiable indicators related to the companies' social, economic and environmental performance and aims to provide stakeholders with measurable information on sustainability issues. Instead of requiring a mere assessment of the reliability of the data provided, the standard requires verifiers to evaluate the extent and nature of compliance with three principles, namely inclusivity, materiality and responsiveness (see below). In this way, the standard sets out to provide a particularly rigorous reporting process (AccountAbility 2008).

Core elements
Inclusivity (stakeholders' participation in developing effective sustainability policies); materiality (evaluating the importance of an issue to the company and its stakeholders); responsiveness (disclosing how the company responds to stakeholder issues that impact on the company's sustainability performance).

Audit
No audit frequency mentioned; third parties/practitioners assure statements' compliance with 15 criteria. AccountAbility independently monitors the compliance of assurance statements uploaded on the CorporateRegister.com website.

Usage to date
No accurate information on AA1000AS is available; 11,801 companies have published and registered their CR reports on the website created and audited by AccountAbility, the drafting entity of the AA1000 series standards (CorporateRegister.com 2015).

Table 4.9 AA1000ES

Summary	
This standard aims to help companies define a clear process to engage their stakeholders in identifying, understanding and responding to sustainability aspects of business conduct.	
Drafting entity	
Institute of Social and Ethical Accountability (ISEA) also known as AccountAbility.	
Released	
2011 (2nd rev.).	
RRI fit	
The rationale behind this standard is that effective engagement of stakeholders confers legitimacy on companies' operations and contributes to their licence to operate. Companies that involve their stakeholders in various aspects of sustainability performance can be better informed about stakeholders' expectations and thus work towards trust-based and transparent stakeholder relationships (AccountAbility 2011).	
Core elements	
Defining stakeholders, purpose and scope of involvement; commitment to inclusivity, materiality and responsiveness (mentioned above under AA1000AS); quality stakeholder engagement process.	
Audit	
A systematic monitoring and evaluation of the overall effectiveness of stakeholder engagement is required. Companies should disclose information on their stakeholder programme and preferably should have their reports independently audited or verified.	
Usage to date	
No accurate information on AA1000ES is available; 11,801 companies have published and registered their CR reports to the website created and audited by AccountAbility, the drafting entity of the AA1000 series standards (CorporateRegister.com 2015).	

4.2.2   Global Initiatives

Global initiatives are mainly developed and published by intergovernmental organizations and come in the form of reporting guidelines or global principles. In contrast with standards, which focus on a single topic, global initiatives have a broader focus and framework for action. They deal with various topics, such as labour relations, environmental management, human rights, consumer protection and fighting corruption (Leipziger 2010). Global initiatives give companies greater flexibility in adjusting their business strategy to initiatives' requirements, and in this sense they are less strict than standards. These CR tools are based on values and principles described in international texts and legislation and have different monitoring mechanisms (De La Cuesta Gonzalez and Martinez 2004). The CR global initiatives selected for our analysis are described in Tables 4.10, 4.11, 4.12, 4.13 and 4.14.

Table 4.10   Global Reporting Initiative (GRI)

Summary
GRI has become synonymous with sustainability reporting as it is the most widely used and promoted tool among companies for reporting results on their social, economic and environmental performance. GRI works closely with the UN Global Compact (see below) to promote principles related to human rights, labour standards, environment and corruption (Russo 2009).
Drafting entity
Coalition for Environmentally Responsible Economies (Ceres), Tellus Institute, United Nations Environment Programme (UNEP).
Released
2013 (4th rev.).
RRI fit
This CR tool fosters the implementation of RRI through its core guidelines. These provide detailed information for companies regarding three main topics namely: (a) the way organizations should report their sustainability performance; (b) the quality of the information provided; and (c) the boundary of the report.[a] GRI also requires companies to report performance on several indicators that cover the following areas: economics, the environment, labour practices and decent work, human rights, society and product responsibility (Russo 2009).
Core elements
Reporting principles for defining content; quality and report boundary; disclosures on: economics, the environment, labour practices and decent work, human rights, society and product responsibility.
Audit
GRI operates at the application level check, a procedure that verifies that a sustainability report has met the reporting requirements. Also, there is a clear tendency nowadays for companies to assure their GRI reports externally. For instance, almost half of the reports listed on GRI's sustainability disclosure database have been given an external assurance.
Usage to date
2,329 (Sustainability Disclosure Database 2015).

[a]Boundary of report: a sustainability report should take into account all entities (e.g. subsidiaries and joint ventures) over which the reporting company exercises control or significant influence both in and through its relationships with various stakeholder groups including suppliers and customers

Table 4.11   UN Global Compact

Summary
The United Nations Global Compact (UNGC) sets a framework for companies to embed in their business strategy ten universal principles in the areas of human rights, labour, environment, and anticorruption.
Drafting entity
United Nations.
Released
2013 (4th rev.).

(continued)

Table 4.11 (continued)

RRI fit
Through its ten principles the UNGC requires companies to improve their performance in areas such as the protection of human rights, respect for differences, the elimination of forced labour, the abolition of child labour, the protection of the environment and the implementation of anti-corruption policies. These topics resulted from a consolidation of several important initiatives, namely the Universal Declaration of Human Rights, the International Labour Organization's Declaration on Fundamental Principles and Rights at Work, the Rio Declaration on Environment and Development and the United Nations Convention against Corruption. The initiative entails a management model which aims to assist companies in implementing the ten principles, measure their performance against these principles and communicate the results through a report (UNGC 2010).

Core elements
Human rights; labour standards; environment; anticorruption.

Audit
In the context of auditing performance, the global compact entails a transparency and accountability policy known as the Communication on Progress (COP). Participating companies are required to publish an annual COP policy in order to demonstrate compliance with the tools' principles.

Usage to date
12,817 (UNGC 2015).

Table 4.12 OECD Guidelines for Multinational Enterprises (MNE)

Summary
These government supported guidelines are mostly recommendations for promoting responsible business conduct among multinational corporations. They cover a wide range of areas aimed at improving the social, environmental and economic performance of the MNE.

Drafting entity
Organisation for economic co-operation and development (OECD).

Released
2011 (5th rev.).

RRI fit
The MNE guidelines promote the implementation of RRI through a principles-based framework, which multinational enterprises need to adopt in order to apply responsible policies. This framework requires companies to disclose information related to their sustainability performance and promotes the implementation of policies that deal with, among other areas, the protection of human rights, compliance with ILO standards, the enhancement of environmental performance, the adoption of anticorruption policies and compliance with taxation requirements, as these are outlined in the OECD Model Tax Convention and the UN Model Double Taxation Convention (OECD 2014).

Core elements
Employment and industrial relations; human rights; environment; information disclosure; combating bribery; consumer interests; science and technology; competition; taxation.

(continued)

Table 4.12 (continued)

Audit
The guidelines entail a proactive agenda that assists multinationals in effectively implementing due diligence, addressing adverse impacts and engaging stakeholders. Also, adhering governments need to set up national contact points, which will be responsible for promoting and monitoring the adoption of the guidelines.

Usage to date
Governments of 44[a] countries have set up national contact points to promote the guidelines and assist in their implementation (OECD 2014).

[a]These are: Argentina, Australia, Austria, Belgium, Brazil, Canada, Chile, Colombia, Czech Republic, Denmark, Egypt, Estonia, Finland, France, Germany, Greece, Hungary, Iceland, Ireland, Israel, Italy, Japan, Korea, Latvia, Lithuania, Luxembourg, Mexico, Morocco, Netherlands, New Zealand, Norway, Peru, Poland, Portugal, Romania, Slovak Republic, Slovenia, Spain, Sweden, Switzerland, Tunisia, Turkey, United Kingdom, USA

Table 4.13 UN Guiding Principles on Business and Human Rights

Summary
These principles are divided into two sections: (a) the state's obligation to respect human rights, and (b) the corporate responsibility to do so.

Drafting entity
United Nations.

Released
2011.

RRI fit
Regarding states, the guidelines outline operating principles that: (a) ensure the enforcement of laws related to human rights; (b) assist businesses in implementing these laws; and (c) require companies to communicate their performance on human rights protection. Regarding the corporate aspect, the tool requires businesses to set a policy that addresses how companies deal with the human rights of their stakeholders. This policy needs to be publicly available and embedded at all levels of business processes (UN 2011).

Core elements
Human rights protection; disclosure; human rights policy.

Audit
As indicated by national and international human rights laws.

Usage to date
No accurate information is available; 140 EU companies have signed agreements with global or European workers' organizations on labour standards (European Commission 2011).

4.2.3 Principles

Principles are CR tools developed mostly by businesses in collaboration with stakeholders such as nongovernmental organizations (NGOs) and trade unions. The topics covered are mostly related to anticorruption, sustainability, and labour rights issues. Principles focus on incorporating the environmental and social dimension into the decision-making processes and strategies of businesses.

Table 4.14 International Labour Organization (ILO) Multinational Enterprises (MNE) Declaration

Summary
The use of the ILO MNE Declaration could, like the previous tools, foster the implementation of RRI in the area of human wellbeing. Its aim is to promote business conduct that ensures economic and social progression. The tool refers not only to companies, but also to governments of both the home and host countries of MNE.
Drafting entity
International Labour Organization (ILO).
Released
2006 (4th rev.).
RRI fit
The tool sets out a framework that companies should adopt in order to ensure compliance with legal requirements, the protection of labour rights and the provision of a healthy and safe working environment. This framework also refers to industrial relations, more precisely to requirements regarding freedom of association and the right to organize as well as collective bargaining.
Core elements
Employment; training; working and living conditions; industrial relations.
Audit
Governments are required to ratify and comply with the major labour conventions. Businesses should consult governments on how to comply with relevant laws.
Usage to date
No accurate information is available; 140 EU companies have signed agreements with global or European workers' organizations on labour standards (European Commission 2011).

Principled CR instruments attribute high importance to the role of commitment by the whole organization and the deep change required in corporate culture so that environmental and social strategies are designed and implemented. Although auditing mechanisms for principles vary, they usually involve independent verification or assurance conducted by third parties. The CR principles selected for our analysis are described in Tables 4.15, 4.16, 4.17, 4.18 and 4.19.

Table 4.15 Business Principles for Countering Bribery

Summary
The development of these principles involves NGOs, companies and trades unions. This tool aims to assist companies effectively when they deal with anticorruption requirements as these are set out in intergovernmental initiatives such as the Global Compact.
Drafting entity
Transparency International.
Released
2013 (3rd rev.).

(continued)

Table 4.15 (continued)

RRI fit
The Business Principles for Countering Bribery require companies to develop an antibribery programme in the context of addressing related ethical challenges and applying a corporate responsibility strategy. They also enjoin companies to disclose information on any political contributions, sponsorships and donations made or received. The principles do not require verification or certification of adoption; rather, they operate as a good practice model and do not involve implementation of all the details they enumerate. Companies are expected to monitor their performance on their own, and only then, if they are interested, seek external verification. To facilitate verification, Transparency International has developed the Assurance Framework for Corporate Anti-Bribery Programmes (Transparency International 2013).

Core elements
Conflicts of interest; bribes; political contributions; charitable contributions and sponsorships; facilitation payments; gifts, hospitality and expenses.

Audit
The Assurance Framework for Corporate Anti-Bribery Programmes strongly encourages users of the Business Principles for Countering Bribery to prepare their public reports according to the International Standard on Assurance Engagements (ISAE3000). The latter defines an independent assurance procedure carried out by a third party.

Usage to date
No data exist as businesses do not 'sign up' as such to the Business Principles.

Table 4.16 Caux Round Table Principles (CRT) for Business

Summary
Executives from Europe, Japan, and the United States were involved in the development of this CR tool. It consists of seven core principles, which are based on three ethical foundations for corporate responsibility and a fair society at large, namely: responsible stewardship; living and working for mutual advantage; and respect and protection of human dignity.

Drafting entity
Caux Round Table.

Released
2009 (2nd rev.).

RRI fit
The CRT principles take risk management into account in order to manage effectively the interests of both businesses and society. Also, the principles are supported by detailed stakeholder management guidelines covering a variety of interested parties including customers, employees, shareholders, suppliers, competitors, and communities (Caux Round Table 2010).

Core elements
Promotion of sustainable development; compliance with the law; support for multilateral trade; support for responsible globalization; respect for the environment; avoidance of illicit operations; stakeholder involvement.

Audit
The tool requires internal audits to ensure that payments made by the company are proper. It also calls for the publication of an annual report that is audited by independent financial or social auditors and suggests collaboration with monitoring organizations such as Transparency International.

Usage to date
No data exist.

Table 4.17   Ceres Roadmap for Sustainability

Summary
This CR tool aims to assist firms in creating a tailor-made programme for effectively managing their social, economic and environmental challenges.

Drafting Entity
Coalition for Environmentally Responsible Economies (Ceres).

Released
2010.

RRI fit
The Ceres roadmap for sustainability sets out 20 requirements for sustainability that interested companies should adopt. These lie broadly in four areas of corporate sustainability, namely governance, stakeholder engagement, disclosure and performance. Companies need to embed sustainability issues into their existing management systems, ensure accountability, engage stakeholders in their strategy and report on how they have dealt with stakeholder input in their everyday activities. Regarding disclosure, the tool requires interested companies to use the GRI guidelines (see above) for reporting. The Ceres roadmap for sustainability also refers to performance and demands that companies: (a) undertake the necessary investment in order to improve their environmental performance; and (b) require their suppliers to meet similar standards. Tangible sustainability requirements are also demanded in the areas of research and development (if applicable) and employee rights (Ceres 2010).

Core elements
Governance for sustainability; stakeholder engagement; disclosure; performance: *operations* (greenhouse gas emissions and energy efficiency; facilities and buildings; water management; waste elimination; human rights); *supply chain* (policies and codes; alignment of procurement practices; engaging suppliers; measurement and disclosure); *transportation and logistics* (transportation management; transportation modes; business travel and commuting); *products and services* (business model innovation; research and development and capital investment; design for sustainability; marketing practices; strategic collaborations); *employees* (recruitment and retention; training and support; promoting sustainable lifestyles).

Audit
The tool requires companies to adopt management systems that entail frequent monitoring, performance assessments and audits. It also demands independent verification of disclosed information by third parties and complaint mechanisms.

Usage to date
The Ceres company network has 63 members (Ceres 2015).

Table 4.18   Ethical Trading Initiative (ETI) Base Code

Summary
This tool is the result of collaboration between companies, NGOs and trades unions and focuses on improving workers' lives by promoting compliance with international labour standards. Companies interested in becoming members of the ETI need to adopt the ETI Base Code and apply practices that respect workers' rights and promote the enhancement of their living conditions.

Drafting entity
Ethical trading initiative.

Released
2012 (The 'decent working hours' clause was revised in 2014).

(continued)

Table 4.18 (continued)

RRI fit
The ETI Base Code requires companies to allow workers the right to collective bargaining and to ensure a healthy and safe working environment as well as employees' integrity and the payment of living wages. Also, the code demands that companies maintain reasonable working hours and not use any child labour or practice discrimination. Moreover, ETI enjoins companies to avoid the excessive use of fixed-term contracts of employment (ETI 2012). In this sense, this CR tool links with RRI in the area of ethical acceptability.

Core elements
Freedom of choice; unions; health and safety; no child labour; living wages; decent working hours; no discrimination; full employment; no physical abuse.

Audit
The implementation of the code is assessed through monitoring and independent verification. To continuously improve the external auditing procedure, participating companies are required to engage with other participants to identify best practices in monitoring and verification and share their experiences.

Usage to date
87 companies (ETI 2014).

Table 4.19 Business Social Compliance Initiative (BSCI)

Summary
This is a business-driven tool focusing on working conditions and aiming at assisting companies in complying with international labour standards.

Drafting entity
Foreign Trade Association (FTA).

Released
2014 (3rd rev.).

RRI fit
Built on ILO conventions and recommendations, this tool sets requirements that protect workers' rights and the environment. More precisely, the tool promotes the implementation of corporate responsibility through the protection of workers' rights to collective bargaining, the prohibition of child labour and of any form of discrimination, and compliance with minimum wage standards and maximum working hours, as these are defined by relevant laws. It also sets requirements that refer to environmental protection, anticorruption, and occupational health and safety. Moreover, this initiative requires participating companies to have a policy for social accountability.

Core elements
Commitment; consistency; comprehensiveness; development-orientation; credibility; focus on risk countries; efficiency; knowledge base; collaboration.

Audit
BSCI audits are conducted only by auditing companies accredited by the international organization Social Accountability Accreditation Services (SAAS). Although this is a non-certifiable tool, participants are encouraged to apply for SA8000 certification. Provision is made for random unannounced checks (RUCs) in factories, and for extra checks on auditors to ensure implementation of the initiative and quality of the audit process.

Usage to date
1,316 companies (BSCI 2015).

4.3 Assessment of Tools

The tools discussed here are all intended to encourage responsible business conduct on an international basis and highlight the following as areas of corporate
responsibility: quality assurance; corporate sustainability; respect for and the
upholding of human rights; workers' rights; the protection of information; and
anticorruption. While in theory these tools are well intentioned, there have been
cases where their effectiveness in managing corporate responsibilities has been
questioned. A point of criticism is that CR tools are voluntary in nature, not legally
enforceable and lack explicit sanctions (King and Toffel 2009). For instance,
though global initiatives like the Global Compact or OECD guidelines offer a
roadmap for companies to engage in corporate responsibility, it is a rather vague
roadmap as it entails neither detailed monitoring requirements nor any sanctions
in the case of violations. Another drawback relates to the fact that there are commercial relationships between companies and auditors, insufficient auditor training and knowledge, and infrequent visits from auditors to certified facilities. The
ambiguity in the tools' content has also been highlighted as a further weakness
(Elis and Keane 2009). For instance, while the ETI Base Code refers to various
aspects of employment and stipulates a number of issues that need to be addressed
in order to enhance working conditions, it includes ambiguities such as an obligation on employers to pay a 'living wage' without a clear definition of what this is.

These issues are no doubt significant, but it would be wrong to claim that all
CR tools are just a fig leaf. There are tools, such as standards, that are designed
to enforce compliance with their requirements by providing detailed guidance
on how to implement them and penalizing firms for misconduct. There are others, such as global initiatives or principles, that might look more abstract, but do
not lack good intentions in promoting responsible business conduct. CR tools deal
with key and complex global challenges, and this is why we need to be careful
when evaluating their effectiveness. Additionally, the task of assessing different
CR tools is further complicated by the fact that they have widely varied primary
objectives.

In search of assessment criteria, we have been inspired by the ISEAL Alliance
(International Social and Environmental Accreditation and Labelling Alliance
Alliance) code of good practice for setting social and environmental tools (ISEAL
2010). ISEAL is a non-profit organization which promotes the use of social and
environmental tools as effective means of making positive social, environmental
and economic impacts. The code includes several criteria that tool-setting organizations need to have in mind when designing their tools. By doing so, tool-set-
ters can ensure that the use of their instruments will bring measurable social and
environmental benefits and will address the aforementioned problems related to
the implementation of CR tools. We have adapted this code to our RRI needs in
Table 4.20.

Having set these criteria, we wanted to see how well the chosen CR tools
performed against them. To do so, we used a five-point Likert scale in which 1

Table 4.20 Tools assessment criteria

Tool development	Tool content and structure
Should be developed in a transparent way.	Follows a logical structure in which objectives are analysed in detail and explicitly linked with required practices.
Must be the result of a wide stakeholder consultation process.	Identifies risks that might threaten the implementation of the tool leading to merely symbolic adoption.
Should engage all stakeholders who: (a) will be directly impacted by the tool; and (b) could influence the implementation of the tool.	Sets clear audit procedures that entail corrective actions and strict sanctions in case of symbolic adoption.
Should not overlap with, but rather complement, other tools.	Entails indicators that can easily be accessed by stakeholders to assess firms' performance.
Must require practices that go beyond compliance with regulatory requirements.	Avoids ambiguous language.
	Demands dissemination of results on firms' CR performance.

represented noncompliance with the assessment criteria and 5 indicated full compliance. There were 11 assessment criteria; hence, a tool that fully complied with all 11 could receive a maximum of 55 points (5×11), whereas a tool that did not meet any could receive no more than 11 points (i.e. 1×11). The total figure we ended up with represents the overall compliance of the tools. Tables 4.21, 4.22 and 4.23 illustrate, respectively, the performance of standards, global initiatives and principles on the assessment criteria.

The tool with the highest score was the Ceres Roadmap for Sustainability, closely followed by GRI. These two CR tools provide a detailed handbook on how to incorporate responsible business conduct into corporate DNA and how to report on CR performance. Interestingly, the most widely promoted CR standard, ISO26000, did not score as highly as expected, mainly due to the fact that it does not incorporate a clear audit procedure (as it is a non-certifiable standard). Also, the two most popular CR tools worldwide, namely ISO9001 and ISO14001, received some of the lowest scores overall. Of the CR tools selected in this book, seven met the criteria at a rate higher than 75 % (with the tool that scored highest achieving an 81 % compliance rate), nine at a rate of 65 % and four at a rate of 55 %. Of the tools that received the highest scores, five were standards, one was a global initiative and one was a principle, whereas the tools with the lowest scores were all global initiatives.

Regarding the performance of tools against each criterion, AA1000AS, AA1000ES, GRI, the Global Compact, the Caux Round Table Principles and the Ceres Roadmap for Sustainability addressed the *stakeholder engagement* criterion most effectively. This criterion is particularly important for RRI as innovation cannot be responsible without taking into account the interests and concerns of those who will be directly or indirectly affected by the development of a new idea.

Table 4.21 CR standards: performance on the assessment criteria

	Criteria	ISO9001	ISO14001	EMAS	ISO50001	OHSAS18001	SA8000	ISO/IEC 27001	ISO26000	AA1000 AS	AA1000 ES
Tool development	Transparency	5	5	5	5	5	5	5	5	5	5
	Stakeholder consultation	5	5	5	5	5	5	5	5	5	5
	Stakeholder engagement	2	3	3	3	3	3	3	3	4	5
	Overlap	5	3	3	5	3	3	5	5	3	4
	Practices beyond regulation	2	3	3	4	3	3	5	5	5	4
Tool content/ structure	Clear objectives linked with practices	3	3	4	4	3	3	4	3	3	3
	Identifying risks	2	2	2	2	2	2	3	2	2	2
	Audit procedure/ sanctions	3	3	3	3	4	3	3	1	3	3
	Indicators	5	5	5	5	5	5	3	3	5	5
	Ambiguous language	2	3	3	4	3	4	3	3	3	3
	Reporting results	1	1	5	2	1	3	4	4	5	5
Total points		35	36	41	42	37	39	43	39	43	44

Table 4.22 CR global initiatives: performance on the assessment criteria

	Criteria	GRI	Global conduct	OECD guidelines for MNE	UN guiding principles on business and human rights	ILO MNE declaration
Tool development	Transparency	5	5	5	5	5
	Stakeholder consultation	5	5	5	5	5
	Stakeholder engagement	4	4	3	2	2
	Overlap	3	3	2	2	2
	Practices beyond regulation	5	3	3	2	2
Tool content/ structure	Clear objectives linked with practices	4	3	3	3	3
	Identifying risks	2	2	2	2	2
	Audit procedure/ sanctions	2	2	2	1	1
	Indicators	5	4	3	3	3
	Ambiguous language	4	3	2	2	2
	Reporting results	5	3	3	3	3
Total points		44	37	33	30	30

Tools with a narrow focus, such as ISO9001, ISO50001, ISO/IEC27001 and ISO26000, received top marks in the *overlap* criterion. The tools that most encouraged companies to *adopt practices beyond regulation* were ISO/IEC27001, ISO26000 and AA1000AS, whereas the tools that scored the highest in the *clear objectives explicitly linked with required practices* criterion, were the Ceres roadmap for sustainability, EMAS, ISO50001, ISO/IEC270001 and GRI.

An interesting aspect of the assessment procedure is the fact that none of the CR tools selected fully acknowledged any risks related with symbolic implementation. Despite its importance, this is an issue that is not currently taken into consideration by tool setters. Also, due to their lack of strict sanctions, none of the CR tools received a score higher than 3 under the audit/sanction criterion.

With reference to the *indicators of performance* criterion, most tools either directly or indirectly referenced the need for setting such measurements; only two principles, Business Principles for Countering Bribery, and the Caux Round Table Principles, made a (relatively poor) reference to such need. Regarding the

Table 4.23 CR principles: performance on the assessment criteria

	Criteria	Business principles for counter-ing bribery	Caux round table principles	Ceres roadmap for sustainability	ETI base code	BSCI
Tool development	Transparency	5	5	5	5	5
	Stakeholder consultation	5	5	5	5	5
	Stakeholder engagement	3	4	5	3	3
	Overlap	4	3	3	3	3
	Practices beyond regulation	1	4	4	2	3
Tool content/ structure	Clear objectives linked with practices	3	2	4	2	3
	Identifying risks	2	2	2	2	2
	Audit procedure/ sanctions	2	2	3	3	4
	Indicators	2	2	5	4	5
	Ambiguous language	3	2	4	3	3
	Reporting results	5	5	5	5	1
Total points		35	36	45	37	35

ambiguous language criterion, the tools that scored the highest were ISO50001, SA8000, GRI and the Ceres Roadmap for Sustainability. These tools did not make excessive use of vague words, and employed language to support their consistent interpretation.

Last but not least, eight tools received top marks for the *reporting* criterion as they require compliant companies to disseminate results on their CR performance. These tools are EMAS, AA1000AS, AA1000ES, GRI, Business Principles for Countering Bribery, the Caux Round Table Principles, the Ceres Roadmap for Sustainability and the ETI Base Code.

4.4 Conclusion

In the current absence of RRI tools for industry, we drew on the vast number of existing CR tools to identify those that link most promisingly with RRI. We developed several criteria to narrow down our choices and select those CR tools with a clear potential to assist practitioners in implementing the RRI principles. This led us to ten CR standards, five CR global initiatives and five CR business principles. The discussion on these tools stressed their fit with RRI and highlighted that they are well suited to achieving at least some of the aims and aspirations of RRI. Equally, the assessment of the tools showed that some are suitable for setting up a framework for action that could form a basis for RRI implementation.

The next chapter develops possible roles for CR tools in RRI and offers some scenarios of what they might look like in practice by mapping key concepts of the RRI discourse onto the CR tools. This discussion will enlighten us further as to which of these tools are suitable for adoption within the context of RRI and which gaps remain.

References

AccountAbility (2008) The AA1000 assurance standard. AccountAbility, London
AccountAbility (2011) The AA1000 stakeholder engagement standard. AccountAbility, London
Alic MA, Rusjan B (2010) Contribution of the ISO9001 internal audit to business performance. Int J Qual Reliab Manage 27(8):916–937
Beschorner T, Müller M (2007) Social standards: toward an active ethical involvement of businesses in developing countries. J Bus Ethics 73(1):11–20
Boiral O (2011) Managing with ISO systems: lessons from practice. Long Range Plan 44(3):197–220
Bondy K, Matten D, Moon J (2008) Multinational corporation codes of conduct: governance tools for corporate social responsibility? Corp Gov: Int Rev 16(4):294–311
BSCI (2015) Participating companies. https://www.bsci-intl.org/about-bsci/members/all?page=48. Accessed 6 Jan 2015
Caux Round Table (2010) CRT principles for responsible business. http://www.cauxroundtable.org/index.cfm?&menuid=8. Accessed 3 March 2014
Ceres (2010) The 21st century corporation: the Ceres roadmap for sustainability. Ceres, Boston
Ceres (2015) Company network members. http://www.ceres.org/company-network/company-directory. Accessed 1 Feb 2015
CorporateRegister.com (2015) Home page. http://www.corporateregister.com/. Accessed 06 Jan 2015
Cragg W (2005) Ethics codes, corporations and the challenge of globalization. Edward Elgar, Cheltenham
Delmas M, Montiel I (2008) The diffusion of voluntary international management standards: responsible care, ISO9000, and ISO14001 in the chemical industry. Policy Stud J 36(1):65–93. doi:10.1111/j.1541-0072.2007.00254.x
Elis K, Keane J (2009) A review of ethical standards and labels: is there a gap in the market for a new 'good for development' label? Working paper 297, ODI, London
ETI (2012) The ETI base code. http://www.ethicaltrade.org/resources/key-eti-resources/eti-base-code. Accessed 13 March 2014
ETI (2014) Our members. http://www.ethicaltrade.org/about-eti/our-members. 6 Jan 2015
European Commission (2011) Communication from the commission to the European parliament, the council, the European economic and social committee and the committee of the region. A renewed EU strategy 2011–14 for corporate social responsibility. European Commission, Brussels

European Commission (2013) Options for strengthening responsible research and innovation. European_Commission. Available via Science with and for Society. http://ec.europa.eu/research/science-society/document_library/pdf_06/options-for-strengthening_en.pdf. Accessed 22 Feb 2014

European Commission (2014a) The EU eco-management and audit scheme. http://ec.europa.eu/environment/emas/about/summary_en.htm. Accessed 22 Feb 2014

European Commission (2014b) EU register of EMAS organisations. http://ec.europa.eu/environment/emas/register/reports/reports.do;jsessionid=UWnfK-_SOAptQhmzHVLxCv-pYA-3WqbjcHirOcjQh-mWLoC87Rml!-2071534248. 12 Jan 2015

Frederick WC (2006) Corporation be good! The story of corporate social responsibility. Dog Ear Publishing, Indianapolis

Gonzalez MDLC, Martinez CV (2004) Fostering corporate social responsibility through public initiative: from the EU to the Spanish case. J Bus Ethics 55(3):275–293

GRI (2013) G4 sustainability reporting guidelines. GRI, Amsterdam

Heras-Saizarbitoria I, Boiral O (2013) ISO9001 and ISO14001: towards a research agenda on management system standards. Int J Manage Rev 15(1):47–65

ISEAL (2010) Setting social and environmental standards v5.0. ISEAL code of good practice. ISEAL Alliance, London

ISO (2009) Environmental management. The ISO14000 family of international standards. ISO, Geneva

ISO (2010) Social responsibility—discovering ISO26000. ISO, Geneva

ISO (2011) Win the energy challenge with ISO50001. Available via ISO. http://www.iso.org/iso/iso_50001_energy.pdf. Accessed 12 March 2014

ISO (2012) ISO and energy. Working for a cleaner sustainable future. Available via ISO. http://www.iso.org/iso/iso_and_energy.pdf. Accessed 8 March 2014

ISO (2013) The ISO survey. http://www.iso.org/iso/iso-survey. Accessed 21 Feb 2015

Jamali D (2010) MNCs and international accountability standards through an institutional lens: evidence of symbolic conformity or decoupling. J Bus Ethics 95(4):617–640

King AA, Toffel MW (2009) Self-regulatory institutions for solving environmental problems: perspectives and contributions from the management literature. In: Delmas M, Young O (eds) Governance for the environment. New perspectives. Cambridge University Press, Cambridge, pp 98–116

Kolesar PJ (1994) What deming told the Japanese in 1950. Qual Manage J 2(1):9–24

Leipziger D (2010) The corporate responsibility code book, 2nd edn. Greenleaf Publisher Ltd., Sheffield

Locke RM, Rissing BA, Pal T (2013) Complements or substitutes? Private codes, state regulation and the enforcement of labour standards in global supply chains. Br J Indus Relat 51(3):519–552

Malsch I, Grinbaum A, Bontems V, Fruelund Anderson AM (2012) Communicating nanoethics .http://ethicschool.nl/_files/Communicatingnanoethicsreportfinal.pdf. Accessed 21 Feb 2014

Nadvi K (2008) Global standards, global governance and the organization of global value chains. J Econ Geogr 8(3):323–343

OECD (2014) National contact points. OECD. http://mneguidelines.oecd.org/ncps/. Accessed 13 March 2014

OHSAS_Project_Group (2011) Results of the survey into the availability of Oh&S standards and certificates, up until 2011-12-31. OHSAS 18001 Expert, London

Runhaar H, Lafferty H (2009) Governing corporate social responsibility: an assessment of the contribution of the UN global compact to CSR strategies in the telecommunications industry. J Bus Ethics 84(4):479–495. doi:10.1007/s10551-008-9720-5

Russo MV (2009) Explaining the impact of ISO14001 on emission performance: a dynamic capabilities perspective on process and learning. Bus Strategy Environ 18(5):307–319. doi:10.1002/bse.587

SAI (2014) SA8000 certified facilities. http://www.saasaccreditation.org/certfacilitieslist.htm. Accessed 6 Jan 2015

SGS (2009) OHSAS18001: 2007 update. http://www.uk.sgs.com/ohsas_18001_2007___update?serviceId=10059735&lobId=19035. Accessed 7 Nov 2009

Sustainability_Disclosure_Database (2015) Global conference on sustainability and reporting NGO round table—GRI reporting statistics. http://database.globalreporting.org/search. Accessed 6 Jan 2015

Toffel MW, Short JL, Ouellet M (2013) Codes in context: how states, markets, and civil society shape adherence to global labor standards. Working paper 13-045, Harvard Business School., Harvard

Transparency International (2013) Business principles for countering bribery. A multi-stakeholder initiative led by transparency international. Transparency International, Berlin

UN (2011) Guiding principles on business and human rights. Implementing the United Nations 'protect, respect and remedy' framework. http://www.ohchr.org/Documents/Publications/GuidingPrinciplesBusinessHR_EN.pdf. Accessed 12 March 2014

UNGC (2010) UN global compact management model. Framework for implementation. http://www.unglobalcompact.org/docs/news_events/9.1_news_archives/2010_06_17/UN_Global_Compact_Management_Model.pdf. Accessed 3 Feb 2014

UNGC (2015) Participants and stakeholders. https://www.unglobalcompact.org/participants/search?utf8=%E2%9C%93&commit=Search&keyword=&joined_after=&joined_before=&business_type=all&sector_id=&listing_status_id=all&cop_status=all&organization_type_id=&commit=Search. Accessed 6 Jan 2015

Vogel D (2010) The private regulation of global corporate conduct: achievement and limitations. Bus Soc 49(1):68–87

Zwetsloot G (2003) From management systems to corporate social responsibility. J Bus Ethics 44(2/3):201–207

Chapter 5
Applying Corporate Responsibility Tools to Responsible Research and Innovation

Abstract Corporate responsibility (CR) tools are mapped onto the von Schomberg, Owen and Science with and for Society (SwafS) definitions of responsible research and innovation (RRI) to see which ones may be suitable for adoption within the context of RRI implementation and which gaps remain. Three case studies show how the CR tools selected can facilitate RRI implementation. The first refers to Abengoa, a highly active company in the field of sustainability; the second discusses Seventh Generation's approach to corporate responsibility and how this links with RRI; and the third presents Teck, a mining company which uses a variety of corporate responsibility tools in managing its operations.

Keywords Responsible research and innovation · Corporate responsibility tools · Governance implementation

In its report on options for strengthening responsible research and innovation (RRI) (European Commission 2013), the expert group on the state of art in Europe on RRI states that a coherent approach to RRI implementation requires the development of a framework that clarifies the criteria, processes and instruments for RRI. The corporate responsibility (CR) tools analysed in the last chapter have an excellent potential to satisfy these recommendations, because they define detailed requirements for safeguarding CR and are based on a process management approach, which can decode the RRI agenda into a set of tangible targets.

This chapter discusses the possible roles of CR tools in RRI and possible RRI practices utilizing these tools. In this context, detailed tables are provided that illustrate the thematic areas of RRI addressed by the CR tools selected. The chapter concludes with three case studies of companies that use a combination of such tools. The intention is to provide additional insights from the corporate world into how the CR tools can be used to satisfy RRI requirements.

© The Author(s) 2016
K. Iatridis and D. Schroeder, *Responsible Research and Innovation in Industry*,
SpringerBriefs in Research and Innovation Governance,
DOI 10.1007/978-3-319-21693-5_5

5.1 Possible Roles of CR Tools in RRI

Our analysis so far demonstrates that CR tools can provide a governance framework that ensures the implementation and advancement of RRI. An interesting aspect of these tools, irrespective of their performance in the assessment criteria, is that they all address widely accepted CR principles that are in line with RRI thematic elements, such as transparency, ethical behaviour, respect for stakeholders' interests, accountability, respect for the rule of law, respect for international norms of behaviour and respect for human rights (ISO 2010).

For instance, standards such as ISO9001, ISO14001, EMAS, SA8000 and OHSAS18001 ensure accountability and transparency as they enjoin certified firms to develop a corporate policy with respect to the issue addressed by each standard, and to make this policy available both internally and externally. In their policies, certified firms need explicitly to state their objectives and responsibilities towards their stakeholders and explain how their operations ensure stakeholder satisfaction (Biazzo and Bernardi 2003; Göbbels and Jonker 2003).

Furthermore, tools like ISO50001 and the Ceres roadmap for sustainability require internal audits to identify weak links in business operations as well as potential opportunities for improvement. They also require external audits that scrutinize firms' operations and call on them to comply with their initial objectives and responsibilities (Ceres 2010; ISO 2011).

In addition, the CR tools selected demand openness about decisions and activities that have an impact on society and the economy, and communication of these in a clear, accurate, timely, honest and complete manner. These tools compel corporate ethical behaviour as their use safeguards quality assurance, corporate sustainability, respect for and the upholding of human rights and workers' rights, the protection of information and protection against corruption. Moreover, they set down communication procedures between the company and its stakeholders, giving the latter insights into how the firm meets their expectations and demands. For example, tools such as AA1000ES and the Caux Round Table Principles for Business (AccountAbility 2011; Caux Round Table 2010) aim to promote stakeholders' participation in developing effective sustainability policies and disclose information on how the company responds to stakeholder issues that impact on the company's sustainability performance.

These CR tools have another important aspect that requires RRI, in that they demand respect for the rule of law and for international norms of behaviour. In particular, they require firms to comply with legal requirements in all jurisdictions in which they operate, and to review their compliance regularly. This is an important aspect of the CR tools, as some companies undertake research and innovation activities without regard to legal requirements (Banerjee 2008). Last but not least, CR tools require respect for human rights and the protection of fundamental principles and rights at work. For instance, in compliance with relevant law, the tools forbid the employment of children and the use of forced or compulsory labour.

What the above discussion illustrates is that CR tools can provide support for research and innovation that is implemented through process development and aligned with the ideas of RRI. For this reason, we argue that CR tools can shape the operationalization of RRI in industry. For instance, RRI could be implemented through the development of a CR plan, which, based on existing corporate responsibility tools, would serve as an indispensable tool of business strategy. In developing such a plan, companies would be able to further their strategic objectives and do well by doing good (Prahalad 2014). Drawing on the literature (Guadamillas-Gomez and Donate-Manzanares 2011) and on tools' requirements, the implementation of RRI practice could be enabled through practices such as:

- internal and external auditing procedures that monitor firms' performance, identify areas for improvement and potential drawbacks, and set up a series of corrective and preventive actions;
- interaction with customers to identify their needs and the development of innovative solutions that satisfy customers' preferences;
- surveys conducted to enhance supply chain management and communication with suppliers;
- the protection of shareholder interests through information transparency, accountability and best practice approaches;
- the dissemination of results on corporate social, environmental and financial performance;
- the acquisition of environmental and quality certification at research centres; and
- the use of systems to promote learning and professional development.

5.2 Possible RRI Practices Using CR

In the preceding section we have argued that CR tools are well suited to achieving at least some of the aims and aspirations of RRI. Here we develop some scenarios of what this might look like in practice by mapping key concepts of the RRI discourse onto the CR tools. These key RRI concepts are drawn from the most widely used definitions in the RRI discourse, namely those of von Schomberg (2013), Owen et al. (2013) and the EU's Science with and for Society programme (2013). The key RRI concepts highlighted in these three definitions are: sustainability; ethical acceptability; societal desirability; risk management related to social, ethical and environmental issues; human wellbeing; anticipation and reflexivity; deliberation (inclusion); responsiveness; open access; gender and science education (see Chap. 2).

The following tables demonstrate that there is an important overlap between CR and established tools to promote both CR and RRI (Tables 5.1, 5.2, 5.3, 5.4, 5.5, 5.6, 5.7, 5.8, 5.9, 5.10, 5.11 and 5.12).

Table 5.1 Sustainability

RRI concept	Sustainability
CR tools	Environmental management standards (ISO14001, EMAS).
	Energy management standards (ISO50001).
	Corporate sustainability standards, global initiatives, principles (AA1000 series standards; GRI; the OECD guidelines for multinational enterprises; the Ceres roadmap for sustainability).
Implementation of the RRI concept	ISO14001 and EMAS enjoin companies to develop a specific methodology for identifying the environmental aspects of their activities and evaluating the environmental impacts stemming from their operations or products.
	ISO50001 helps companies conserve resources and tackle climate change at large.
	GRI requires companies to report performance on several indicators that cover the following areas: economics, the environment, labour practices and decent work, human rights, society and product responsibility.
	The Ceres roadmap for sustainability calls on companies to embed sustainability issues in their production operations.

Sustainability involves the critical use of natural resources and investment in eco-efficient methods of production. The CR tools include detailed guidance on how to identify the environmental aspects of business operations, enhance the use of natural resources and invest in environmental performance improvements

Table 5.2 Ethical acceptability

RRI concept	Ethical acceptability
CR tools	Quality management standard (ISO9001).
	Standards and principles focusing on health and safety and on workers' rights (OHSAS18001, SA8000; Ethical Trading Initiative (ETI) Base Code).
	Corporate social responsibility standard ISO26000.
	Global initiatives and principles focusing on respect for and the upholding of human rights (UN Global Compact, UN Guiding Principles on Business and Human Rights, ILO MNE Declaration, ETI Base Code).
Implementation of the RRI concept	ISO9001 requires the company not to practise price gouging, make misleading advertising claims or sell ineffective, unreliable or unsafe products.
	OHSAS18001 and SA8000 require the use of safe equipment that does not threat employees' integrity at all stages of the company's operations.
	ISO26000 requires, among other things, ethical corporate conduct.
	The ETI Base Code requires companies to apply practices that respect workers' rights and promote the enhancement of their living conditions.
	The UN guiding principles refer to state and corporate obligations to respect human rights.

By employing CR tools for the purposes of RRI, companies may be able to bring to the fore discussions of ethical acceptability and thereby come to a shared understanding. Ethical acceptability furthermore has a bearing on the question of the financial aspect of innovation. This aspect can take the role of either a barrier or an enabler, depending on whether or not RRI is perceived to have long-term financial benefits. Ethical acceptability, however, is a broader concept than financial viability, and the relationship between acceptability and financial viability is likely to be a potential area of conflicting values

Table 5.3 Societal desirability

RRI concept	Societal desirability
CR tools	Environmental management standards (ISO14001, EMAS).
	Energy management standards (ISO50001).
	Standards and principles focusing on health and safety and on workers' rights (OHSAS18001, SA8000, Ethical Trading Initiative (ETI) Base Code).
	Corporate social responsibility standard (ISO26000).
Implementation of the RRI concept	ISO14001 and EMAS contribute towards protecting the rights of local communities through the application of pollution control measures.
	ISO50001 contributes in the battle against greenhouse gas emissions and, by extension, in the battle against global warming.
	OHSAS18001, SA8000 and the ETI Base Code contribute towards better working environments where employees' integrity is ensured and the use of children as a workforce is prohibited.
	ISO26000 enjoins, among other things, respect for stakeholders' interests.

Societal desirability has a similar status to ethical acceptability in our analysis, in that it is not an explicit theme but implicitly linked to some, such as values, including financial ones

Table 5.4 Risk management related to social, ethical and environmental issues

RRI concept	Risk management related to social, ethical and environmental issues
CR tools	Environmental management standards (ISO14001, EMAS).
	Energy management standards (ISO50001)
	Standards and principles focusing on health and safety and on workers' rights (OHSAS18001, SA8000, the Ethical Trading Initiative (ETI) Base Code).
	Information security management standards (ISO27001).
	Anti-corruption global initiatives and principles (UN Global Compact, Business Principles for Countering Bribery, the Caux Round Table principles).
Implementation of the RRI concept	ISO14001 and EMAS require the identification of any environmental risks and the implementation of measures for the mitigation of those risks
	OHSAS18001, SA8000 and the ETI base code reduce the risks associated with health and safety at work by (a) identifying the health and safety aspects of an organization's activities; and (b) minimizing the risk of accidents and any legal violations.
	ISO27001 reduces information security risks.
	The Caux Round Table principles take into account risk management in order to effectively manage the interests of both businesses and society.

Risk management is most closely related to the prediction of future consequences. A potential problem here is that risks are typically understood in a very specific manner, namely as the financial values of potential damages multiplied by their probability. This view of risk allows for successful risk management, but it struggles to deal with risks that are not quantifiable or whose probabilities are unknown. The CR tools such as the Caux Round Table principles aim to move beyond this limited view of risk, but it is not clear to what degree they can capture the often very fuzzy risks that can arise from research and innovation activities

The Tables are Coming below that many CR tools map onto the core concepts of RRI, but not onto all of them. More precisely, the thematic topics of gender and science education cannot be fully addressed by the tools (e.g. science education is only partially addressed by one tool, namely the OECD Guidelines for

Table 5.5 Human wellbeing

RRI concept	Human wellbeing
CR tools	Global initiatives and principles focusing on respect and upholding human rights (UN Global Compact, UN Guiding Principles on Business and Human Rights, ILO MNE Declaration, the ETI Base Code).
Implementation of the RRI concept	These tools require the implementation of various measures and policies for the protection of human rights and the promotion and enhancement of living conditions.

Human wellbeing is core to RRI, but it is difficult to ascertain which activities are conducive to wellbeing and which are not. CR tools provide a list with thematic areas that offer a starting point for narrowing down policies that contribute to human wellbeing. Examples include, but are not limited to: ensuring health and safety in the workplace; no child labour; paying employees a living wage; ensuring decent working hours; no discrimination; ensuring full-employment; no physical abuse; collective bargaining; and freedom of choice

Table 5.6 Anticipation

RRI concept	Anticipation
CR tools	Quality management standard (ISO9001).
	Environmental management standards (ISO14001, EMAS).
	Energy management standards (ISO50001).
	Corporate sustainability standards, principles and global initiatives (AA1000 series standards, the CERES Roadmap for Sustainability, GRI, the OECD Guidelines for Multinational Enterprises).
	Standards and principles focusing on health and safety and on workers' rights (OHSAS18001, SA8000, the Ethical Trading Initiative (ETI) Base Code.
	Information security management standards (ISO270001).
Implementation of the RRI concept	These CR tools are audited and verified under a clear and systematic procedure. During audits, firms can identify areas that need improvement and implement corrective actions. Also, during audits, firms can anticipate potentially negative impacts related to their business operations and implement preventative actions to mitigate those impacts.
	The ETI Base Code in particular, requires participating companies to engage with other participants in order to share their experiences and identify best practices in monitoring.

RRI is predicated on the idea that it is possible, at least to some degree, to anticipate future outcomes of research and innovation and consequences (both intended and non-intended) for broader groups. The CR tools listed here may go some way towards broadening the scope of organizational anticipation and addressing consequences for a range of stakeholder groups such as employees, customers, suppliers and government

Multinational Enterprises). Nevertheless, the tools address the vast majority of the RRI thematic elements, and in many instances these elements are covered not by a single tool, but by a variety, illustrating that major RRI topics are deeply rooted in major CR tools.

Table 5.7 Reflexivity

RRI concept	Reflexivity
CR tools	Quality management standard (ISO9001). Environmental management standards (ISO14001, EMAS). Energy management standards (ISO50001). Corporate sustainability standards, principles, global initiatives (AA1000 series standards, the Ceres Roadmap for Sustainability; GRI; the OECD Guidelines for Multinational Enterprises).
Implementation of the RRI concept	ISO standards require companies to review their performance, at least annually, and during this review to set targets, objectives and indicators for the next year. In this way, companies can revise their activities, commitments and assumptions, identify strengths and weaknesses and, if needed, set corrective or preventative actions.
	Corporate sustainability standards and principles require companies to publish results on their performance on social, environmental and economic aspects of their business operations. In this way, companies can review their performance against these areas and evaluate the level of commitment they achieved, in relation to previous years.

Reflexivity is a difficult to grasp concept and linked to a longstanding discourse in the social sciences. From an RRI perspective it would seem to imply that not only the means of achieving particular research and innovation goals, but also the ends that are to be achieved, are reflected upon. To ensure that this happens, auditing and reviewing mechanisms are needed. Through these mechanisms, companies can assess their performance, identify areas for improvement and set new targets. The CR tools cover such procedures and can provide detailed guidance into how to implement such actions

Table 5.8 Deliberation (inclusion)

RRI concept	Deliberation (inclusion)
CR tools	Corporate social responsibility standard (ISO26000).
	Quality management standard (ISO9001).
	Environmental management standards (ISO14001, EMAS).
	Energy management standards (ISO50001).
	Corporate sustainability standards, principles and global initiatives (AA1000 series standards, the Ceres Roadmap for Sustainability, GRI, the OECD Guidelines for Multinational Enterprises).
	Standards and principles focusing on health and safety and on workers' rights (OHSAS18001, SA8000, the Ethical Trading Initiative (ETI) Base Code.
	Information security management standards (ISO27001).
	Anticorruption global initiatives and principles (UN Global Compact, Business Principles for Countering Bribery, the Caux Round Table Principles).
	Global initiatives and principles focusing on respect and upholding human rights (UN Global Compact, UN Guiding Principles on Business and Human Rights, ILO MNE Declaration, the ETI Base Code).
Implementation of the RRI concept	These tools are the outcome of a wide consultation process, so that interests of both internal and external stakeholders are taken into account.
	The AA1000 series standards in particular require companies to include stakeholders in the development of effective sustainability policies, and to evaluate the importance of an issue to the company and its stakeholders.

Deliberation or inclusion is covered by stakeholder involvement. What is less clear is to what degree the inclusion of broader stakeholder groups plays a role and can be relevant to RRI in industry. The RRI literature strongly suggests a broad set of criteria for the selection of stakeholders. The CR tools may help companies to come to such a broader understanding of stakeholders and define ways of engaging them successfully

Table 5.9 Responsiveness

RRI concept	Responsiveness
CR tools	Environmental management standards (EMAS).
	Corporate sustainability standards, principles and global initiatives (AA1000 series standards, the Ceres Roadmap for Sustainability, GRI, the OECD Guidelines for Multinational Enterprises).
Implementation of the RRI concept	The tools mentioned here require companies to disclose how the company responds to stakeholder issues that impact on the company's sustainability performance.

Responsiveness is most closely related to stakeholder involvement. It is furthermore driven by the company ethos. Companies are typically responsive to those stakeholders whom they perceive as important and legitimate. Broadening this group and ensuring responsiveness to other stakeholders is in the spirit of RRI and may be promoted by CR tools

Table 5.10 Open access

RRI concept	Open access
CR tools	Environmental management standards (EMAS).
	Corporate sustainability standards, principles and global initiatives (AA1000 series standards, the Ceres Roadmap for Sustainability, GRI, the OECD Guidelines for Multinational Enterprises).
Implementation of the RRI concept	The tools mentioned here require companies to disclose how the company responds to stakeholder issues that impact on the company's sustainability performance.

Open access refers to publishing results of the research and innovation process free of charge. Several CR tools require companies to report on their performance and upload their corporate reports on their websites, so that these are available to everyone, free of charge. However, this does not include the disclosure of research and innovation results

Table 5.11 Gender

RRI concept	Gender
CR tools	Standards and principles focusing on health and safety and on workers' rights [SA8000, the Ethical Trading Initiative (ETI) Base Code, Business Social Compliance Initiative (BSCI)].
	Global initiatives and principles focusing on respect and upholding human rights (UN Global Compact, UN Guiding Principles on Business and Human Rights, ILO MNE Declaration).
Implementation of the RRI concept	These tools are against any discrimination based on: gender, race, caste, national origin, religion, disability, sexual orientation, union membership, political affiliation or age.
	The CR tools focusing on human rights indirectly refer to this topic too. For instance, the UN Global Compact requires maternity protection.

Gender refers to ensuring gender balance in decision-making and in research teams. Although this is not directly addressed by CR tools, some dealing with health and safety in the workplace or with workers' rights are against any form of discrimination, including gender

Table 5.12 Science education

RRI concept	Science education
CR tools	Corporate sustainability global initiatives (the OECD Guidelines for Multinational Enterprises).
Implementation of the RRI concept	This CR tool deals with this topic to a degree as it recommends the development of commercial objectives in line with universities and public research institutions, in order to foster the diffusion of research and development activities.

Science education refers to innovative ways of connecting science with society, so that science becomes more attractive to young people, for instance. To achieve this, teaching and learning processes are important to familiarize young people with science and raise their awareness of the different aspects of science and technology. However, these are not included explicitly in CR tools

5.3 Case Studies

To demonstrate how the CR tools selected can pave the way for RRI implementation, we discuss three cases of companies that use some of the CR tools analysed to address their corporate responsibilities. As will be shown, these companies do not focus on a single CR tool, but use a combination of tools to integrate corporate responsibility into their business strategy and promote a way of doing business that addresses several of the RRI thematic areas.

5.3.1 Abengoa

Abengoa (www.abengoa.com) is a highly innovative company that uses technological developments to provide sustainability solutions. The company focuses on the energy and environment sectors, and specializes in the generation of electricity from renewable resources, the conversion of biomass into biofuels and the production of drinking water from seawater. Innovation is a key component of the company's business model. Abengoa operates several research and development centres to improve the efficiency of its current innovative solutions for sustainability and to develop new solutions such as marine energy. With reference to the latter, the company collaborates with international technology partners to produce its own technology by generating energy from sea waves (marine energy) (Abengoa 2014).

Corporate responsibility is highly important for the company, which uses several CR tools to address its corporate responsibilities. The firm has built a CR management system in line with ISO26000, has adopted the UN Global Compact and has created a computer-based integrated sustainability management system. The latter comprises subsystems such as greenhouse gas, environmental management, and health and safety management systems. In 2014 Abengoa, in line with several CR standards such as ISO14001, ISO14067, ISO50001, ISO26000, SA8000, GRI G4 and OHSAS18001, completed the development of an internal norm that will enable the company to calculate its global footprint. Also, to ensure that the company achieves its CR aspirations and implements the CR management system, it has established five norms that are compulsory for all Abengoa staff (Box 5.1).

Box 5.1 Abengoa's norms on CR (adapted from Abengoa 2014)

- **Norm on CR** (mapping onto the RRI principles of ethical acceptability, societal desirability, anticipation, reflexivity and responsiveness): the norm relates to the implementation of the CR system, its management, auditing and reporting.
- **Norm on quality and environmental management** (mapping onto the RRI principle of risk management related to social, ethical and environmental issues): the norm focuses on issues such as customer concerns and environmental aspects of the firm's operations and how these are addressed.
- **Norm on human resources** (mapping onto the RRI human wellbeing principle): the norm aims to ensure a fair working environment for the firm's employees and covers human rights, diversity, equality, training and occupational risk.
- **Norm on management of legal affairs, risk analysis and insurance management** (mapping onto the RRI principle of risk management related to social, ethical and environmental issues): the norm focuses on corporate governance, risk management (including sustainability risks) and legal aspects of the firm's operations such as contracts with suppliers and partners.
- **Norm on consolidation, auditing and management of fiscal affairs** (mapping onto the RRI anticipation and reflexivity principles): the norm deals with anticorruption, auditing, and internal control and transparency.

The norms presented in Box 5.1 provide a solid basis for the continuous implementation of the CR system and a good indication of how firms can put in practice key RRI concepts. In particular, the norm on human resources addresses the topic of human wellbeing, whereas the norm on the consolidation, auditing and management of fiscal affairs deals with the anticipation thematic area. In turn, the norm on legal affairs management, risk analysis and insurance management addresses the risk management topic, while the norm on CR ensures that the company effectively satisfies its corporate responsibilities in line with societal expectations; hence, the norm partially addresses societal desirability.

Of particular interest are Abengoa's code of social responsibility (based on the principles of the UN Global Compact and the SA8000 standard); its labour social responsibility policy and management system implementation (based on the SA8000 standard); and its universal risk model to manage risks related to human rights violations effectively either in the company's facilities or in supplier activities.

To put on record its progress in the social, environmental and financial areas, the firm publishes a corporate social responsibility report in line with the GRI principles and the AA1000AS standard. This report also provides evidence of how the company is meeting the UN Global Compact requirements. The report's reliability is ensured through external evaluation by GRI and verification by an external verifier in the context of the AA1000AS standard. These two tools are useful in the context of RRI and can provide important guidance on RRI implementation as they require the development of a CR report that addresses certain RRI principles.

- First, GRI and AA1000AS require stakeholder engagement and inclusivity. To take into account stakeholders' views and develop sustainability policies in line with their preferences and concerns, Abengoa conducted interviews with 22 stakeholder representatives.
- Second, both tools require a firm to clarify how it meets stakeholders' concerns related to its sustainability performance. Abengoa has developed guidelines and an integrated sustainability management system to achieve the incorporation of those concerns into its business strategy. The firm also provides constantly updated information to its stakeholders so that they can evaluate company performance.
- Third, by publishing its CR report the firm assesses its sustainability performance and communicates its results to its stakeholders, thus raising their awareness of the topic. The published report allows stakeholders to verify the firm's progress as it provides comparable information, written in an accurate and clear way and made available on time, and thus make informed decisions about the firm's performance. In this sense, the GRI and AA1000AS requirements satisfy the sustainability and partly, the open access principles of RRI.

5.3.2 Seventh Generation

Seventh Generation (www.seventhgeneration.com) is another case of a highly innovative company that uses CR tools to implement a responsible way of innovating and doing business in general. Sustainability is a foundational principle for the company, which has been focused since its inception on safeguarding the needs of future generations. The company sells cleaning, paper, and personal care products generated with human and environmental wellbeing in mind (Seventh Generation 2014).

The tools the company uses offer further insights into how CR tools can facilitate the implementation of RRI in industry. In particular, Seventh Generation participates in the Ceres network of companies and works on four strategic areas highlighted by the Ceres Roadmap for Sustainability, which also address several RRI thematic elements (see Box 5.2). These areas are governance, stakeholder engagement, disclosure and performance.

Box 5.2 CR areas overlapping with RRI principles

- **Governance** (mapping onto the RRI sustainability principle): key activities are board of directors oversight for sustainability; management accountability; executive compensation links to environmental, social and governance performance; and the strength of sustainability policies and management systems.
- **Stakeholder engagement** (mapping onto the RRI inclusion principle): the company commits to engaging in continuous dialogue with its stakeholders across the whole value chain, and to taking into account stakeholder feedback in decision-making.
- **Disclosure** (mapping onto the RRI responsiveness principle): the company is committed to reporting regularly on its sustainability performance and to providing stakeholders with credible metrics on that performance as well as detailed information for future actions.
- **Performance** (mapping onto the RRI sustainability, ethical acceptability and social desirability principles): the company undertakes investments to improve its environmental performance and insists that its suppliers meet the same social and environmental standards.

More precisely, to integrate sustainability into its business model and ensure the conservation of natural resources, Seventh Generation uses in its products materials based on plants that are grown and harvested responsibly, and not petroleum-based materials. Also, to minimize waste the firm uses recycled materials and biodegradable or recyclable packaging. Furthermore, Seventh Generation created the Companies for Safer Chemicals organization to promote the reform of state and federal regulations on toxic substances, so that environmental protection and sustainability are safeguarded and effectively addressed in legislation.

The company claims that, in pursuit of human and environmental wellbeing, its products do not endanger consumers' health and do not damage the environment. Examples include non-chlorine-bleached, 100 %-recycled paper towels, tissues, and napkins; nontoxic, phosphate-free cleaning products; recycled plastic trash bags; and chlorine-free baby and feminine care products.

Seventh Generation engages with a number of stakeholder groups such as employees, consumers, suppliers, retailers, Ceres, the Environmental Working

Group, B Corp (B Lab), Greenpeace, the Rainforest Alliance, the Forest Stewardship Council, Healthy Families, the Breast Cancer Fund and American Sustainable Business. To address stakeholder concerns more effectively, the company seeks stakeholder input in a multistep process. This involves surveys, employee and consumer feedback, insights provided during a stakeholder review organized by Ceres, input from internal teams such as the company's corporate consciousness team and executive leadership team, and feedback from Ceres on the company's report. This feedback has helped the company define its key priorities with reference to corporate responsibility. In this context, Seventh Generation prioritizes respect for local communities, securing a safe and vibrant working environment for its employees, ensuring that the company's supply chain conforms to the same social and environmental standards and promoting chemical legislation that will result in safer consumer products.

To inform its stakeholders about its actions Seventh Generation reports on its performance in line with GRI requirements; these provide a detailed framework for ensuring that the information provided is credible, understandable, comparable and easily accessible. In its report the company uses the GRI G4 guidelines to provide information about its performance in several thematic areas such as stakeholder engagement, governance, ethics and integrity, the environment, labour practices and product responsibility. Stakeholders can access the information free of charge from the company's website (hence satisfying the open-access thematic requirement of RRI) and get information on the company's priorities and performance.

5.3.3 Teck

Teck (www.teck.com) is a mining company specializing in a range of activities, namely exploration, development, mining and minerals processing, safety, environmental protection, materials stewardship, recycling and research (Teck 2014). The company is a major producer of copper, seaborne steelmaking coal and zinc, while it also invests in energy efficiency. Teck pays particular attention to its corporate responsibilities and constantly increases its efforts to integrate corporate sustainability into its business strategy. Indicative of its commitment to CR is the fact that the company has been named in the Dow Jones Index World Index for the fifth year in row.

The firm's corporate responsibility programme has six areas, namely community, people, water, biodiversity, energy, and materials stewardship. To monitor progress in these areas the company has set objectives, targets and indicators that focus on the short (2015) and long term (2030). Critical to the firm's success in promoting a responsible way of doing business and achieving its sustainability goals have been the various CR tools the company implements in its daily routines. In particular, the firm's CR policy and actions are constantly informed by a range of CR tools including the AccountAbility (AA) 1000 Standards, the Global

Reporting Initiative (GRI), the International Labour Organization (ILO) standards, ISO14000 and ISO26000, and the United Nations Guiding Principles on Business and Human Rights.

These tools assist Teck in setting a framework for implementing corporate responsibility as they provide the company with a precise methodology for the identification and effective management of corporate responsibility issues. In this context, the company has developed a health, safety, environment and community management system (Box 5.3) to promote a consistent approach in dealing with its CR programme focus areas (Teck 2014).

Box 5.3 Policy documents of Teck's health, safety, environment and community management system (adapted from Teck 2014)

- **Charter of Corporate Responsibility** (mapping onto the RRI principles of ethical acceptability, social desirability, anticipation, reflexivity and responsiveness): Teck has developed here a set of topics referring to safeguarding business ethics, health, the safety environment and community wellbeing in all of its operations.
- **Code of Sustainable Conduct** (mapping onto the RRI sustainability principle): This document explains how the company's operations enhance community and corporate environmental performance.
- **Health and Safety Policy** (mapping onto the RRI principles of human wellbeing, risk management related to social, ethical and environmental issues, and responsiveness): This presents the company's commitment to providing leadership and resources for entrenching the core value of safety.
- **Human Rights Policy** (mapping onto the RRI human wellbeing and responsiveness principles): This policy presents the company's commitment to respecting the rights of its employees, local communities and others affected by the firm's operations.

The way Teck addresses the first two of its key corporate responsibility areas, i.e. community and people, provides useful insights into how to implement the RRI topics of human wellbeing, anticipation, reflexivity, inclusion and gender in a corporate context.

Regarding community, Teck conducted social impact assessments and social risks analyses at a number of its operations, developed and negotiated three comprehensive agreements with indigenous people, and designed and implemented feedback mechanisms at several of its operations. Furthermore, the firm built partnerships with other private companies and with international organizations such as UNICEF to develop a zinc and health programme that has improved the lives of thousands of people in sub-Saharan Africa and India.

In the context of the key area of people, Teck has identified nine categories of stakeholders: employees, local communities, society at large, government, indigenous peoples, media, shareholders, business partners and industry associations. The company has set priority engagement topics for each stakeholder group. For instance, for its employees such topics include, but are not limited to, career and professional development, as well as health and wellbeing, while for society at large such topics are community investments and mining practices and activities. Also, Teck pays particular attention to ensuring a working environment that is free from any form of discrimination, and to do so it has developed and implemented antidiscrimination and antiharassment policies.

Moreover, the company's activities in the areas of water, biodiversity, energy and materials stewardship provide interesting insights into the implementation of the RRI areas of sustainability, risk management and societal desirability. More precisely, Teck's policy on water aims to safeguard water quality, the efficient use of water in its operations and the fair allocation of water between the company and its local communities. To achieve that, the firm has set a water balance that applies to all of its operations and monitors the volumes of water consumed. It has also designed and implemented integrated water management plans for its operations and collaborated with local communities to produce a water quality plan.

With reference to biodiversity, Teck has developed biodiversity management plans at some of its operations and purchased land for wildlife and habitat conservation purposes. The company's endeavours have been recognized by the local community: the firm received an award for its reclamation activities at the Pinchi Lake mine in British Columbia, Canada.

Teck has also been very active in the area of energy consumption, having invested in the implementation of an anti-idling programme for haul trucks at mining operations, which reduced the firm's greenhouse gas emissions by 13,000 tonnes of carbon dioxide and saved approximately five million litres of diesel fuel. Likewise, the company has developed and implemented energy reduction projects that have reduced energy consumption by 650 terajoules since 2011. To improve its performance in product stewardship, the firm has continued to pressure its suppliers and partners for performance in line with the company's responsible practices as these are informed by the various CR tools.

The company also commits itself to communicating with stakeholders in an accurate, understandable and comparable manner through its corporate responsibility report. The report is prepared in line with GRI requirements and independently reviewed by an external verifier. This also demonstrates how businesses can address the RRI topic of responsiveness.

5.4 Conclusion

This chapter has mapped key concepts of the RRI discourse onto CR tools and demonstrated that major RRI elements are inherent to these tools. Thus we have illustrated that the RRI thematic areas can be addressed by several tools, including the tools that scored highly in the assessment criteria discussed in the previous chapter (see Chap. 4) and those that did not. This is a significant finding as it proves that CR tools in general can be used for implementing RRI. Obviously, the tools that received top marks in the previous chapter have greater potential to assist firms in implementing RRI principles, but, as was shown, all CR tools can provide useful guidance on this topic. Hence, these tools can help interested parties overcome major problems stemming from the new and fluid nature of the RRI concept. Likewise, the various detailed requirements found in CR tools can narrow down RRI from an abstract concept into a set of tangible targets.

Furthermore, the case studies have provided real-life insights into how a combination of CR tools enables firms to integrate corporate responsibilities into business strategy and promote business conduct that is in line with RRI principles. The next chapter deepens knowledge by presenting a list of systematic questions on RRI and CR.

References

Abengoa (2014) Annual report 2014. 02 corporate social responsibility. http://www.abengoa. com/export/sites/abengoa_corp/resources/pdf/en/gobierno_corporativo/informes_anuales/201 3/Tomo2/2013_Volume2_AR.pdf. Accessed 12 Feb 2015
AccountAbility (2011) The AA1000 stakeholder engagement standard. Accountability, London
Banerjee SB (2008) Corporate social responsibility: the good, the bad and the ugly. Crit Sociol 34(1):51–79. doi:10.1177/0896920507084623
Biazzo S, Bernardi G (2003) Process management practices and quality systems standards: risks and opportunities of the new ISO9001 certification. Bus Process Manage J 9(2):149–169
Caux Round Table (2010) CRT principles for responsible business. http://www.cauxroundtable.org/ index.cfm?&menuid=8. Accessed 3 March 2014
Ceres (2010) The 21st century corporation: the Ceres roadmap for sustainability. Ceres, Boston
European Commission (2013) Options for strenghtening responsible research and innovation. In: European Commission: available via science with and for society. http://ec.europa. eu/research/science-society/document_library/pdf_06/options-for-strengthening-en.pdf. Accessed 22 Feb 2014
Göbbels M, Jonker J (2003) AA1000 and SA8000 compared: a systematic comparison of contemporary accountability standards. Manage Audit J 18(1):54–58
Guadamillas-Gomez F, Donate-Manzanares MJ (2011) Ethics and corporate social responsibility integrated into knowledge management and innovation technology. J Manage Dev 30(6):569–581
ISO (2010) Social responsibility: discovering ISO26000. ISO, Geneva
ISO (2011) Win the energy challenge with ISO 50001. Available via ISO. http://www.iso.org/iso/ iso_50001_energy.pdf. Accessed 12 March 2014
Owen R, Stilgoe J, Macnaghten P, Gorman M, Fisher E, Guston D (2013) A framework for responsible innovation. In: Owen R, Bessant J, Heintz M (eds) In responsible innovation. Wiley, London, pp 27–50
Prahalad CK (2014) The fortune at the bottom of the pyramid. Pearson, New Jersey
Science with and for Society (2013) Responsible research and innovation. http://ec.europa.eu/ programmes/horizon2020/en/h2020-section/science-and-society. Accessed 22 Feb 2015
Seventh Generation (2014) A generation of good: 2013 corporate consciousness report. http://generationofgood.7genreport.com/. Accessed 30 Jan 2015

Teck (2014) Teck 2013 sustainability report. http://www.tecksustainability.com/sites/base/pages/performance-data-pages/report-archive. Accessed 15 Feb 2015

von Schomberg R (2013) A vision of responsible research and innovation. In: Owen R, Bessant J, Heintz M (eds) In responsible innovation. Wiley, London, pp 51–74

Chapter 6
'Are You RRI-Aware?' a Question-and-Answer Chapter for Innovators

Abstract The book ends with a dialogic tool that asks the reader questions about responsible research and innovation (RRI), responsibility, corporate responsibility, risk assessment, diversity, user involvement and other aspects of RRI to test awareness of research and innovation governance concepts and tools. Sample answers are provided.

Keywords Responsible research and innovation · Corporate responsibility

In conclusion, we summarize the book and its main suggestions through questions and answers. For our suggested answers, please turn to Sect. 6.3.

6.1 Introduction

Those who are part of a larger business are likely to have done some personal development or legal training through e-learning. For instance, you may have been briefed on new anticorruption laws and how to implement them; or you may have gone through an e-training session on antidiscrimination before being involved in staff recruitment. At the end of the e-training, you are likely to have found a question and answer section where you collected points, or your percentage of correct answers was recorded. If you scored 100 %, everybody was happy. If you scored below a certain threshold (e.g. 80 %), you had to redo the training.

This is not that kind of question and answer session. We are not going to test how well you have remembered the contents of this book, and there are no scores. Emerging concepts cannot be captured with multiple-choice options. Even though we have significant expertise in industry and business,[1] we hope that this chapter

---

[1]Before becoming an academic, Konstantinos Iatridis worked as a business consultant and implemented CR management systems in more than 60 companies, some of them in the FTSE 100, while Doris Schroeder worked as a budget planner for Warner Music Manufacturing Europe.

© The Author(s) 2016   83
K. Iatridis and D. Schroeder, *Responsible Research and Innovation in Industry*,
SpringerBriefs in Research and Innovation Governance,
DOI 10.1007/978-3-319-21693-5_6

will be read and pondered by *current* innovators who have their own well-informed views. While the next section is designed to reinforce our arguments, views and analysis, we acknowledge that RRI is still an emerging concept and that there are no definitely right or wrong answers (apart from obviously wrong ones, such as 'pollution is desirable').

6.2 Questions

To know whether one's activities fall under a certain umbrella, one needs to know what they are. Hence, the first question is obvious.

What is RRI?

Once it is relatively clear what is meant by a concept or framework or tool, the main question is about the incentive for complying with it.

Why should innovators be interested in RRI?

The core of the RRI concept is 'responsibility'.

6.2.1 Responsibility

What does 'responsibility' mean?

Have disagreements about moral responsibilities been discussed in your company?

Has the core of any potential problem in your company ever been defined as 'ethical acceptability'? What areas does 'ethical acceptability' cover, in your view?

Or was the core of the problem 'sustainability'? What areas does 'sustainability' cover in your view?

Is your company interested in offering products and services for the benefit of society? And how would you define 'societal desirability'?

Having understood what responsibility means does not equate with knowing how to realize or implement its goals.

6.2.2 How to Discharge RRI Responsibilities

Does your company operate a risk management system? What is its main focus? Is it about environmental risks?

Does your company actively work towards achieving a diverse workforce? What measures does it take to increase diversity?

Can staff members at all levels input feedback and ideas into the decision-making process? Do they know about the engagement model, if you have one?

Can end users of your products and services input feed back and ideas into the decision-making process? How do they know about your end-user engagement model, if you have one?

Can other stakeholders in your products and services input feedback and ideas into the decision-making process? How do they know about your engagement model, if you have one?

Some tools to discharge responsibilities are available.

6.2.3 Corporate Responsibility Tools

How do you manage corporate responsibility issues such as quality, environmental protection, health and safety and employee protection in your company?

Is your CR management approach based on any CR tools? If so, have you been certified by any of these tools?

Do you undertake any practices that go beyond mere compliance with regulatory requirements?

Have you set clear objectives for corporate responsibility aims that are linked with practices?

Do you monitor your business operations regularly?

Does your monitoring involve a formal procedure or an informal one?

Do you report your performance results against corporate responsibility targets?

And how well are you doing?

6.2.4 RRI Self-Assessment

> ### Do you believe your company operates responsibly?

> ### Who drives responsible decision-making in your company?

> ### Can you see a difference between operating legally and operating responsibly? If so, what is the difference?

6.2.5 About You and Your Company

Question	Answer
Position in company	
Year of company's establishment	
Number of employees	
Industry	
Number of persons involved in corporate responsibility management	
Country/ies where your company operates	

6.2.6 Are You RRI-Aware?

If you have read this far, and worked through the questions, it is likely that you are interested in corporate responsibility and RRI. The following section discusses the kind of answers one could give.

6.3 Suggested Answers and Guidelines

What is RRI?

The concept is still under development, but RRI covers three areas:

1. *Ethical acceptability* is governed by legal instruments and ethics guidelines. The most straightforward examples, and possibly the oldest, are the requirements placed on researchers and innovators in medical research. For instance, they may not test any products on human participants who have not given their consent.
2. *Sustainability* is a very broad concept, describing the requirement that research and innovation should meet the needs of the present without compromising the ability of future generations to meet their own needs. Efforts to achieve sustainability are driven by legal instruments (e.g. the Cartagena Protocol on Biosafety), environmental certification tools (e.g. ISO14001 on environmental management) and personal efforts.
3. *Societal desirability* is the least explored of the three RRI areas. It requires research and innovation to have the potential to benefit humankind as a whole, and also to address the research and innovation needs of marginalized and impoverished populations.

These areas represent the three goals or ambitions of RRI. There are various tools that can achieve their fulfilment, including:
- public, stakeholder and end-user engagement;
- measures to achieve a balanced workforce, e.g. antidiscrimination measures; and
- the use of existing corporate responsibility tools, for instance, to decrease negative environmental impact.

Why should innovators be interested in RRI?

In a market economy it would be ideal if the main incentive for RRI interest or compliance were increased profitability. However, while we have given examples (e.g. the ambiact, see Chap. 2) of market entry being accelerated and facilitated through RRI measures, this principle does not hold universally.

A more honest answer is that reflective practitioners of any profession, including industry researchers and innovators, are likely to care about how they are viewed from outside as well as inside their industry. Are they responsible practitioners in the eyes of their peers? Are they responsible practitioners in the eyes of the public? Are they themselves satisfied with their own conduct?

What does 'responsibility' mean?

In litigation environments, responsibility means the obligation to answer for an act done (or left undone), and the expectation that the defendant will repair any injury caused. In the context of RRI, responsibility is more pro-active and inclusive. It asks: what more can you do to conduct your company's business responsibly? Especially, what moral obligations do you have? And how can you discharge them all coherently?

Moral obligations are often the most complex. Philosophers have been debating moral obligations of various kinds for thousands of years, and corporate responsibility obligations only entered the frame in the 20th century. RRI obligations have emerged only in the 21st century.

The 'newspaper and mother' test has been recommended to check whether you are fulfilling your moral obligations in terms of business ethics. The test asks: 'Would you want your conduct to be published in a respected broadsheet, and read about by your mother?' While this is a good general test, it has its shortcomings. For instance, a CEO who closes down a research team due to long-term nonprofitability and to safeguard the interests of the remaining research staff might read in the newspaper one morning that some of the researchers have slid into serious depression or alcoholism. 'The point here is that sometimes even ethically good decisions are ones that we wouldn't want publicized … because their negative consequences are more visible than their positive ones' (MacDonald 2010).

There are no obvious short-cuts to ascertaining the moral obligations of researchers, innovators and their managers. Reflective initiative within given contexts is required, enhanced by consultation with peers and stakeholders.

Have disagreements about moral responsibilities been discussed in your company?

Such moral disagreements could, for instance, be about where research and innovation funds are invested, with some controversial emerging technologies being the most likely to attract disagreements.

Has the core of any potential problem in your company ever been defined as 'ethical acceptability'? What areas does 'ethical acceptability' cover, in your view?

As noted in an earlier answer, ethical acceptability is governed by legal instruments and ethics guidelines. The most straightforward examples, and possibly the oldest, are the requirements placed on researchers and innovators in medical research. For instance, they may not test any products on human participants who have not given their consent.

Or was the core of the problem 'sustainability'? What areas does 'sustainability' cover in your view?

As noted in an earlier answer, sustainability is a very broad concept, describing the requirement that research and innovation should meet the needs of the present without compromising the ability of future generations to meet their own needs. Efforts to achieve sustainability are driven by legal instruments (e.g. the Cartagena Protocol on Biosafety), environmental certification tools (e.g. ISO14001 on environmental management), and personal efforts.

Is your company interested in offering products and services for the benefit of society? How would you define 'societal desirability'?

As noted in an earlier answer, societal desirability is the least explored of the three RRI areas. It requires research and innovation to have the potential to benefit humankind as a whole, and also to address the research and innovation needs of marginalized and impoverished populations.

Does your company operate a risk management system? What is its main focus? Is it about environmental risks?

Risk management systems are an integral part of the majority of larger businesses today. While most focus on profitability, or employee health and safety risks, some also focus on broader risks, e.g. risks to the environment.

Does your company actively work towards achieving a diverse workforce? What measures does it take to increase diversity?

Some measures to increase diversity in the workforce can be legal requirements. In the UK, for instance, the Equality Act of 2010 tries to ensure that nobody is discriminated against on the basis of age, disability, gender reassignment, marriage and civil partnership, pregnancy and maternity, race, religion or belief, sex or sexual orientation (Equality Act 2010: sec.4). However, whether this goal is achieved or not depends on the effectiveness of the mechanisms put in place to do so. And anti-discrimination laws on their own are not enough to achieve a diverse workforce. You may have other non-compulsory mechanisms in place.

Can staff members at all levels input feedback and ideas into the decision-making process? Do they know about the engagement model, if you have one?

In large companies, for instance, are staff given opportunities such as question-and-answer sessions with managers, surveys with narrative elements or dedicated emails collecting views? Is there a recognized trade union?

Can end users of your products and services input feedback and ideas into the decision-making process? How do they know about your end-user engagement model, if you have one?

For instance, are civil society organizations that represent end users invited to get involved in product and service development?

Can other stakeholders of your products and services input feedback and ideas into the managerial decision-making process? How do they know about your engagement model, if you have one?

Examples of stakeholders not yet mentioned are government, the general public, suppliers and shareholders.

How do you manage corporate responsibility issues such as quality, environmental protection, health and safety and employee protection in your company?

CR management might be undertaken through a number of procedures such as: quality monitoring of production; product specifications; product identification and traceability; storage, packaging, loading and transport of products; quality plans; identification of environmental aspects and environmental impact assessment; waste management (toxic or nontoxic solid or liquid); air pollution management; noise control; hazard identification; risk assessment and determining controls; health and safety; freedom of association and the right to collective bargaining; discrimination; working hours.

Is your CR management approach based on any CR tools? If so, have you been certified by any of these tools?

Examples of such tools might include ISO90001, ISO14001, EMAS, ISO27001 and OHSAS18001 (See Chap. 4).

Do you undertake any practices that go beyond mere compliance with regulatory requirements?

Such practices might include environmental investments in topics not addressed by relevant legislation; avoiding the use of legal but potentially harmful or controversial ingredients; including stakeholders' views in decision-making; not taking advantage of information asymmetries; ensuring transparency in the production process.

Have you set clear objectives for corporate responsibility aims that are linked with practices?

Examples of an objective, target and indicator might be:
- Objective: Increase recycling levels for next year.
- Target: By 20%.
- Indicator: Tonnes of recycled material throughout the year.

- Objective: Improve employees' awareness of responsible research and innovation.
- Target: By 2016.
- Indicator: Number of training seminars employees attend on responsible research and innovation.

Do you monitor your business operations regularly?

Most CR tools that involve an external certification procedure require at least one annual external audit. However, companies that are truly interested in gaining the benefits of such tools will integrate the audit procedure into their business operations and conduct several internal audits each year. Apart from the benefits of timely identification of any deviations from the tools' requirements, and keeping the tool 'alive', documented internal audits are used during external audits as a proof of the tool's integration into everyday business operations.

Does your monitoring involve a formal procedure or an informal one?

A formal procedure is likely to involve verification by a third party, or making formal reports public so that they can be verified by third parties. For instance, the formal requirement of the UN Global Compact is that participating companies post an annual communication on progress (CoP) report (UN Global Compact 2015).

Do you report your performance results against corporate responsibility targets?

A gradually increasing number of companies report their performance by using various CR tools. Examples might be EMAS, AA1000 AS/ES, Business Principles for Countering Bribery, the Caux Round Table Principles, the Ceres Roadmap for Sustainability and the ETI Base Code.

Such issues might include compliance with a tool's requirements, security risks, environmental aspects, issues related to quality, employee protection, customer satisfaction and customer complaints, communication of CR policy throughout the company.

Do you believe your company operates responsibly?

This is your personal opinion.

Who drives responsible decision-making in your company?

You might think, for instance, that responsible decision-making in your company is driven by external forces though legal requirements. Or there might be internal staff members or groups who drive the process.

Can you see a difference between operating legally and operating responsibly? If so, what is the difference?

Stakeholder engagement, for example, is only very rarely required legally (it could be legally required, say, for mining companies). This would be a difference between operating legally and operating responsibly.

You have now completed reading the book. Thank you. In addition, we are very happy to provide personal feedback on your individual answers and thoughts, if you email them to us (ki267@bath.ac.uk and dschroeder@uclan.ac.uk).

References

Equality Act (2010) Chapter 15. http://www.legislation.gov.uk/ukpga/2010/15/contents. Accessed 24 May 2015

MacDonald C (2010) Business ethics and the "New York Times" rule, 8 Dec 2010. http://business ethicsblog.com/2010/12/08/business-ethics-and-the-new-york-times-rule/. Accessed 24 May 2015

UN Global Compact (2015) What is a COP? https://www.unglobalcompact.org/COP/index.html. Accessed 24 May 2015

About the Authors

Dr. Konstantinos Iatridis is a Lecturer in Business and Society in the School of Management at the University of Bath. He has worked with FTSE 100 listed companies on a range of corporate responsibility/sustainability consulting projects and completed a Ph.D. in Corporate Responsibility and an M.Sc. in Environmental Management culminating in over 14 years of academic and practical experience in this area.

Prof. Doris Schroeder was educated in Germany and the United Kingdom at postgraduate level in management and economics as well as politics and philosophy. Her first career was in financial planning for Time Warner. Her voluntary work includes 15 years for amnesty international. She currently holds two appointments. She is Director of the Centre for Professional Ethics at the University of Central Lancashire, College of Health and Professor of Moral Philosophy, Law School, UCLan Cyprus. She is also an Adjunct Professor at the Centre for Applied Philosophy and Public Ethics at Charles Sturt University, Canberra, Australia. Schroeder is best known for her work on benefit sharing and global justice and has published two earlier Springer books (*Indigenous Peoples, Consent and Benefit Sharing*, co-edited with Rachel Wynberg and Roger Chennells and *Benefit Sharing—From Biodiversity to Human Genetics*, co-edited with Julie Cook Lucas).

© The Author(s) 2016
K. Iatridis and D. Schroeder, *Responsible Research and Innovation in Industry*,
SpringerBriefs in Research and Innovation Governance,
DOI 10.1007/978-3-319-21693-5

Index

A

AA1000 series, 41, 42, 68, 70–72
AA1000AS, 58, 61, 75
AA1000ES, 58, 61, 66
Abengoa, 73–75
Accountability, 11, 33, 41, 42, 66–68, 76, 77
Act, 14, 41, 42
Actions, 6, 8, 10, 25, 35–37, 42, 43, 67, 76, 77
Ambiact, 15, 16, 25
Ambiguous language, 61
Anticipation, 23, 67, 70, 74, 78
Anticorruption, 35, 52, 57, 71, 74, 83
Assessment criteria, 57, 58, 66
Audit, 42, 53, 58, 60, 66, 70
Auditing mechanisms, 53
Auditors, 57
Auditor training, 57
Audit procedures, 67

B

Bottom of the pyramid, 19, 20, 27
Business case, 36
Business conduct, 36, 39, 57, 58
Business impacts, 35
Business principles, 60, 61, 69, 71
Business strategy, 32, 34, 36, 39, 49, 67, 73,
    75, 77

C

Care, 1, 3, 11, 16, 21, 23–25, 34, 75, 76
Caux round table, 58, 60, 61, 66, 69, 71
Ceres, 58, 60, 61, 66, 68, 70–72, 76
Ceres roadmap, 61, 68, 72, 76
Certification, 41, 42, 67

Certification bodies, 42, 43, 48
Charity, 33
Check, 42
Civil society organisations, 13, 15, 25
Code of conduct, 36
Colgate-Palmolive, 36, 37
Collective bargaining, 70
Commitment, 23, 24, 32, 37, 53, 71, 77, 78
Communication procedures, 66
Competitors, 42
Compliance, 10, 13, 15, 35, 36, 57, 58, 66, 72
Contractual responsibility, 10, 12, 16
Core elements, 43–56
Corporate accountability, 33
Corporate citizenship, 33
Corporate culture, 53
Corporate philanthropy, 3, 17, 25, 32, 33
Corporate social performance, 33, 34
Corporate sustainability, 33, 57, 66, 68, 70–72,
    77
Corrective, 42, 67, 70, 71
Corruption, 7, 49, 66
Countering bribery, 60, 69, 71
Customers, 32, 42, 67, 70

D

Data protection, 36
Decision-making, 11, 52, 72, 76
Deliberation, 23, 67, 71
Deming cycle, 41
Development, 2, 6, 14, 16, 17, 33, 39, 42, 58,
    65, 67, 73–75, 77, 79, 83
Discrimination, 7, 26, 35, 70, 72, 79, 83
Do, 2, 3, 17, 25, 35, 42
Documentation, 42, 43

© The Author(s) 2016   101
K. Iatridis and D. Schroeder, *Responsible Research and Innovation in Industry*,
SpringerBriefs in Research and Innovation Governance,
DOI 10.1007/978-3-319-21693-5